T0321437

Ergodesign Methodology
for Product Design

Ergodesign Methodology for Product Design

A Human-Centered Approach

Marcelo M. Soares

CRC Press
Taylor & Francis Group
Boca Raton London New York

CRC Press is an imprint of the
Taylor & Francis Group, an **informa** business

Panton chair cover image kindly provided by Verner Panton Design.

1st edition published 2022
by CRC Press
6000 Broken Sound Parkway NW, Suite 300, Boca Raton, FL 33487-2742

and by CRC Press
2 Park Square, Milton Park, Abingdon, Oxon, OX14 4RN

ISBN: 9781032054483 (hbk)
ISBN: 9781032103273 (pbk)
ISBN: 9781003214793 (ebk)

DOI: 10.1201/9781003214793

Typeset in Times
by Deanta Global Publishing Services, Chennai, India

I dedicate this book:

To my dear teacher and friend Prof. Anamaria de Moraes (in memoria). She was my first source of inspiration in ergonomics studies. I am very grateful for their encouragement since the beginning of my academic career. "Anamaria forever".

To my mother, Maria Teresa Soares, for all her incommensurable effort in prioritizing education and my formation. I owe her everything I am. I am eternally grateful.

Contents

Preface

The challenge in design is to provide users of products with what they really want. Thus, matching customer needs with product characteristics is crucial. Customers are those best able to express their own needs.

Soares (2012) argues that the paradigm of human-centered design as applied to products and systems is to improve user satisfaction and efficiency of use, increase comfort, and assure safety in normal use and in the foreseeable misuse of a product or a system. Achieving these design goals is a significant challenge. The use of appropriate design methods contributes to improving the critical usability features of products and workplaces, such as the ease with which it can be used, its functions learned, and its efficiency, comfort, safety, and adaptability has taken advantage of, such that all of these meet users' needs and contribute to user satisfaction.

Rebelo et al. (2012) state that, when developing a product, the user experience should be part of a User-Centered Design methodology. In this context, it is necessary to clarify that Ergonomics has as its precept the analysis, methodology, and intervention directed to the human being. As such, it presents similar objectives to user-centered design in order to provide users with good product use experiences, meeting their needs, skills, and limitations that are materialized in positive emotions.

Good experiences in using the product are treated as UX (user experience). In order to have a good experience, good usability, good interface, and good ergonomics are necessary. The user interface is the intermediary between the user and the technology or system. We can have from graphical user interfaces (GUI), which are the traditional computer screens, to brain controlled interfaces (BCI), multimodal interfaces (MMI), and ubiquitous interfaces (UI). In short, the interface is the mediator between the user and the technological system. The concepts of user-centered design, usability, good interface, and good user experience have always been part of the ergonomic principles. There is no ergonomics without a design (project), without a human being (user), and without a good interface/optimization of the human–machine system (good user/worker experience). Therefore, such concepts are consonant with each other, thus contributing to the enjoyable use of users with the product and its environment.

We are pleased to present the book *Ergodesign Methodology for Product Design: a Human-centered Approach*. The purpose of this book is to introduce an ergonomic design methodology in which the "user's voice" can be translated into product requirements in a way that designers and manufacturers can use. This characterizes it as a co-design methodology.

It is important to present the difference between **"Human-Centered Design" (HCD)** and **"User-Centered Design" (UCD)**. Although the difference between the two terms is quite subtle, Yalanska (2021) argues that HCD "is the process of creating things based on general natural characteristics and peculiarities of human psychology and perception". UCD, on the other hand, according to the author, is more specific. It focuses not only on human characteristics and perception in general, but also on the specific characteristics of the target users to make the solution of problems directed to meeting the needs of its users. UCD is more related to the way users do things. We adopted in our methodology the term "human-centered approach" because it is more generic and incorporates the aspects related to human emotion and perception, in our view essential to provide user satisfaction.

This book is based on a revised and updated version of the author's doctoral thesis, entitled "Translating user needs into product design for people with disabilities: a study of wheelchairs" (Soares, 1999), defended at Loughborough University, England. We clarify that despite the time in which this methodology was originally conceived, it is, in essence, timeless, and the revision and updating that has been carried out has made it more aligned with the market and technology news.

Although wheelchairs were chosen as the product for study, the methodology presented may be used with any consumer product. First, a review of the literature on ergonomics and product design is presented. Then there is a discussion of issues including consumer needs, product requirements, and user satisfaction. Finally, a human-centered design methodology based mainly on the findings of the literature review is presented.

In this book, we have adopted the term **Ergodesign** as a methodology in which the knowledge of ergonomics is applied to design practice. In this term, Ergodesign has the objective of producing products, machines, tools, or a workplace adapted to the demands of the user, operator, worker, guaranteeing their pleasure, comfort, safety, satisfaction, quality of life, and increased productivity in the workplace.

This book is intended for designers, architects, and engineers involved in product design and development and undergraduate and graduate students in ergonomics, design, architecture, engineering, and related fields. It can also be used by students and professionals of physiotherapy and occupational therapy interested in designing products for people with special needs.

I hope that from the methodology proposed here, we can contribute to the design of safer and more pleasant consumer products that meet the needs of users and promote satisfaction in use.

Prof. Marcelo M. Soares, Ph.D.
September 2021
Acknowledgments or Credits List (<u>unless supplied as end/back matter and clearly labeled as such</u>, applies to certain lists)

Acknowledgments

This book would not have existed without the support of several friends and colleagues who have contributed at different times to its realization.

To friends Aaron Marcus, Claudia Mont'Alvão, Eduardo Ferro, Ernesto Filgueiras, Fabio Campos, Frida Marina Fischer, Gaela Vilela, Gabriela Cuenca, Laura Martins, Manuela Quaresma, Marcio Alves Marçal, Martin Maguire, Maurício Duque, Tareq Ahram, Pradip Kumar Ray, Rosalio Avila Chaurand, and Symone Miguez for their suggestions and revision of the text.

To my son Gabriel Soares and my daughter-in-law Luisa Barros Correia for their support in making some images.

To Prof. He Renke and Prof. Tom Zhao for the constant support and encouragement during my stay at the School of Design, Hunan University, China.

To Prof. Stuart Kirk (in memoria) for the competent guidance in my doctorate even when his health insisted on not helping.

My deep acknowledgments for the support of the School of Design, University of Hunan, China, the Department of Design, Federal University of Pernambuco, Brazil, and the Brazilian agency CNPq (National Council for Scientific and Technological Development) for their support.

Author Biography

 **Marcelo M. Soares** is currently a Full Professor at the School of Design, Hunan University, China, selected for this post under the 1000 Talents Plan of the Chinese Government. He is also a licensed Full Professor of the Department of Design at the Federal University of Pernambuco, Brazil. He holds an M.S. (1990) in Industrial Engineering from the Federal University of Rio de Janeiro, Brazil, and a Ph.D. from Loughborough University, England. He was a post-doctoral fellow at the Industrial Engineering and Management System Department, University of Central Florida. He served as an invited lecturer at the University of Guadalajara, Mexico, University of Central Florida, USA, and the Technical University of Lisbon, Portugal. Dr Soares is Professional Certified Ergonomist from the Brazilian Ergonomics Association, where he was president for seven years. He has provided leadership in ergonomics in Latin American and in the world as a member of the Executive Committee of the International Ergonomics Association (IEA). Dr Soares served as Chairman of IEA 2012 (the Triennial Congresses of the International Ergonomics Association), held in Brazil. Professor Soares is currently a member of the editorial board of *Theoretical Issues in Ergonomics Science* and several journal publications in Brazil. He has about 50 papers published in journals, over 147 conference proceedings papers, and 25 books, and 83 book chapters. He has undertaken research and consultancy work for several companies in Brazil. Prof. Soares is co-editor of the *Handbook of Human Factors and Ergonomics in Consumer Product Design*, the *Handbook of Usability and User-Experience (UX)*, and the *Handbook of Standards and Guidelines in Human Factors and Ergonomics* published by CRC Press. His research, teaching, and consulting activities focus on manufacturing ergonomics, usability engineering, consumer product design, information ergonomics, and applications of virtual reality and neuroscience in products and systems. He also studies user emotions when using products and techniques in real and virtual environments based on biofeedback (electroencephalography and infrared thermography).

1 Introduction

This introduction begins by presenting a hypothetical situation of a disabled user and her difficulties in using a consumer product. Everyday products are presented, and possible failures, misuse, and accidents in handling such products are discussed. It is also discussed the designers' commitment to consider in the design of their products the users' needs, abilities, and limitations, contemplating their diversity in physical and cognitive aspects. The importance of developing products for all segments of the population that focus on user needs should be a priority area in the product design process. This will be the focus to be presented in the following chapters. It appears that the "voice of the users with or without disability" is not being heard by designers. In light of this, it is proposed to investigate (a) the relationship between user needs and product requirements from an ergonomics point-of-view; (b) the current methods that designers use to design the product; and (c) the views of users on the product they use and what demands they make in the design. Finally, the introduction presents a number of issues involving, for example, the role of users, ergonomics, safety, standard, and product design in product development, the design for the disabled people, the involvement of users in the product design, and the production of a methodology in which the voice of the user can be considered in the various stages of project development. These questions will be answered in the following chapters of the book.

The brief fictitious story below allows this introductory chapter to highlight some aspects of the relationship between products and their use that will be covered in the rest of this book. Although starring by a disabled user, this story can be applied to any consumer product user, disabled or not.

Mrs. Cindy is in her late sixties and is considered a modern lady. Although she has a moderate disability, she has used a wheelchair since her childhood. She looks after her home and husband and brings up her young son. Mrs. Cindy lives in a very comfortable apartment, but has a number of complaints concerning her wheelchair, the objects, and environments surrounding her, as reported in the following statement.

"In a typical morning of my daily life, she says, I wake up at about half-past six in the morning. I make breakfast using a fancy new microwave oven that my husband gave me as a present a couple of weeks ago. It was meant to do everything you could ever want to do with a microwave, but it is too complicated to use. My husband (who is a doctor) said he wouldn't go near it because of the difficulty in using it. Fortunately, I've memorized some settings and just ignore the rest.

After doing a bit of housework, like doing some washing, playing with my grandson, and sometimes doing some gardening, I go out to the playground, shopping, or visit some friends or relatives. It is quite hard to manage get my grandson in the back seat,

DOI: 10.1201/9781003214793-1

fasten his seat belt while I am using my wheelchair. I also have problems transferring myself from my wheelchair to the driving seat. The wheelchair is quite large and heavy and doesn't give me enough mobility.

I have 15-hours of assistance a week. A carer helps me with shopping and housework, but I am absolutely convinced that this weekly cost could be avoided if I could manage most of the domestic activities myself using appropriate equipment and user-friendly products. Unfortunately, I live in a very unfriendly environment. My district is very hilly and has virtually no dropped curbs. The shops, local Leisure Centre and other public places where I used to go on a daily/weekly basis have restricted spaces, steep slopes, and cambers or steps which block the way. Neither of the two manuals or the powered wheelchair that I've got can overcome these problems easily. I usually need the help of somebody else.

None of my wheelchairs have the height of the seat adjustable. So, I have difficulty seeing and reaching things at most levels when I am shopping, doing housework, and playing with my toddler grandson. I can't get my powered wheelchair into my car because it is very heavy and cumbersome. So, it is also difficult to carry and control my grandson and shopping from the manual wheelchair because I have to use both my hands for propulsion. My husband complains of pain in his back when pushing my chair because he is very tall, and the wheelchair push-handles are not adjustable. So, the chair forces him to bend forward when pushing me.

I am definitely not very happy with my chairs. I would like a fashionable chair. A chair that represents my personality with a nice design, including bright colors. One that when people see me will make them think 'there's a lady who can do things herself and if she needs help, she'll ask for it'. With my grey, ugly, cumbersome chairs I find people are always saying 'do you need help with this or do you need help with that'. I find this so frustrating. If I need help, I'll ask for it. I need a powerful wheelchair that should be lightweight, highly maneuverable, reliable, and which permits me to cope with lifts, corridors, crowded places, and the boot of my car. These are not the characteristics of the wheelchairs that I actually own".

This fictitious story has all the ingredients of reality and corresponds with the experiences of many, if not most, disabled people. Unfortunately, it also corresponds with the experiences of able-bodied people using ordinary consumer products as well. Whether products are designed for the able-bodied or the disabled, or both, they should be useful in a functional, pleasurable, and safe way. This is not what is happening in an extremely vast number of cases.

If we consider the products used in our daily activities, it is not hard to find those which fail to satisfy our needs, e.g. the correct programming of a microwave oven or the setting up of a smartphone; the use of the remote control of a new and sophisticated smart TV set; the adjustment of self-service photocopier machines, and so forth. Dealing with such products may often result in error and frustration. Additionally, design failures in everyday products make a considerable contribution to the number of accidents at home. According to the data from the *Home and Community Overview* (Injury Facts, 2019), over the last ten years, home and community deaths have increased by 60%, and the death rate per 100,000 people has increased by 49%, with a cost of $472.6 billion in the United States. According to the United States' Injury Facts (2019) website, home and community deaths include all preventable injuries that are not work related and that do not involve motor vehicles on highways. Thus, most of these deaths involve the handling of consumer products.

NOTE

It is estimated that, in the last 10 years, almost 38 million people sought medical assistance related to injuries from home and community deaths, which included accidents with consumer products, in the last ten years in the United States (Injury Facts, 2019).

Nowadays, a vast number of consumer products have reached a level of complexity and difficulty, which is usually not well accepted by users. Although the degree of technological sophistication has produced a strong appeal from the point of view of market strategy, it can produce serious frustration for the user. Most of these times, the products lack maturity and the functional content that the user really needs. Users are no longer satisfied with products that meet only technological criteria, they desire products that they can use in a safe, efficient, comfortable, and pleasurable way.

It is broadly known that designers usually design products with some presumptions about the consumers' expectations, and how they will behave with the products. The designers usually believe that the products suitable for themselves will be equally suitable for others. Consequently, such presumptions usually consider that the users of everyday objects are healthy adults, in very good perceptual, cognitive, emotional, and physical conditions.

Designers who use the above-mentioned approach (and the quality of the products in the marketplace indicates that a great number of them design that way) will probably fail twice. Firstly, they will fail because they are experts in the use of the products that they themselves design, forgetting to consider the needs, abilities and requirements of a representative sample of users. Secondly, they will fail because they also forget that in addition to an extremely diverse population of consumers – in terms of physical and cognitive capabilities – there are a considerable number of people whose physical and mental capabilities are quite diverse from the level of the majority of the population. Not considering the largest range of users' requirements in the design of the product is to condemn the users with limited capabilities to have difficulties which can lead to failure, misuse, and accidents. In fact, if consumer products – which are almost exclusively designed to be used by the standard of the majority of the population instead of the population with fully physical and mental capabilities – are responsible for a large number of home accidents, what happens when these products are used by those with lower levels of physical and/or mental skills?

Products can be designed to meet the needs of a broader spectrum of users without diminishing their value. In fact, the design of products suitable for the great majority, irrespective of age, sex, or physical ability, is a question of respect for human dignity. In a truly inclusive society, the human–machine interface must be such that first and foremost, it will not damage the user's health, but will also respect diversities in the same way that correct town-planning eliminates structural barriers (Dahlin et al., 1994) and provide inclusive products and services (Diversity & Inclusion in Tech, 2018).

INTERNET

Some companies have become aware that designing consumer products including for people with disabilities is a good marketing and social responsibility strategy. Apple introduces a number of functions on the iPhone that make the life of the disabled user more accessible. Watch the link below.

Technology is most powerful when it empowers everyone. (2020)
https://www.apple.com/accessibility/

Although disabled users may have diminished sensory or motor capabilities, limited cognitive ability, or emotional difficulties, their needs are, in general, similar to those of the able-bodied population. So, apart from the needs related to their own disabilities, disabled users have needs in terms of aspirations, uniqueness, values, and status which should be reflected in the products that they use. Dissatisfactions in using the product will occur if the products do not fully meet their needs (e.g. Mrs. Cindy's new microwave oven in the fictional story in the beginning of this chapter). This applies to both kinds of products: those products for general use of the entire population and those designed to meet the needs of the disabled in particular.

The freedom to choose, one of the most precious – and fragile – human quality, is responsible for the sense of independence. Independence depends on choice. The quality of life that people experience as they perform daily activities relates directly to the number and types of choices available. This, of course, depends critically on economic and social, conditions. When people, especially the disabled, are faced with hostile physical environmental conditions and unfriendly products and furniture in their homes that limit choices, there will be frustrations, and a reduction in pleasure, independence, and quality of life.

Disabled people have difficulties using those consumer products designed for the general use of the majority of the population. As previously mentioned, a significant number of those products do not perform their functions, as expected, when in use by able-bodied users. They are not also designed considering the part of the population which has limited physical and/or mental abilities. To overcome these difficulties, disabled people sometimes carry out adaptations to enable these products to meet their needs. So, the everyday products used by the disabled population have been designed following two approaches: (a) by adaptation of existing products and the development of special aids and (b) by taking into account the limitations and capabilities of the disabled in the design of new products.

Products designed specifically for disabled people, as will be discussed in detail in the next chapter, frequently derive from a medical perspective. This limited approach does not consider a number of aspects such as people's aspirations, uniqueness, values, status, and lifestyle, which are regularly considered in the design of products for the able-bodied population. Many of these products stigmatize the user and often increase the user's sense of disability and dependence. In view of this, many products designed for the disabled may be rejected and abandoned even though they are of clinical benefit.

The importance of developing products for all segments of the population that focus on user needs should be a priority area in the product design process. So, to match user needs with product characteristics is the first, and maybe the most important, phase during the product development process. The user's role involves more than simple consultation, but rather includes using the user as a partner in the design and development process (Collaborating with Customers in Product Development, 2007; Swallow, 2018; Gardiner and Rothwell, 1985). Considering direct and indirect users as partners in the design process is one of the principles of co-design. Unlike participatory design, which involves users only on a consultative basis, co-design presupposes their participation in design decisions (Casali, 2021; Cavignau-Bros and Cristal, 2020).

REFLECT

One fundamental question arises from this point. How can the user needs be translated into specifications – particularly ergonomic specifications – in the several phases of product development?

Undoubtedly, users are the best people to say what their needs are.

But it appears that the "voice of the users with or without disability" is not being heard by designers. If the assumption that the "voice of the user" is not being heard is confirmed, there will be a need to investigate:

- the relationship between user needs and product requirements from an ergonomics point of view;
- the current methods that designers use to design the product; and
- the views of users on the product they use and what demands they make in design.

The Chartered Institute of Ergonomics and Human Factors (2019, page 4), previously Ergonomic Society, UK, points out that the "user-centered design is both a design philosophy and a design process. As a philosophy, it makes the needs, wants, and limitations of the end user of a product the priority focus, and as a process it offers designers a range of methods and techniques to ensure this focus is sustained through the various stages of design". We will use this approach in this book.

INTERNET

Watch this video in which Don Norman discusses the practice of human-centered design. In this book we understand that human-centered design is synonymous with user-centered design.

The Changing Role of the Designer: Practical Human-Centered Design, Don Norman, 5 June 2020.
https://www.youtube.com/watch?time_continue=1&v=QewRjNfG1-8&feature=emb_logo

Improving wheelchair design still remains a very current need. Besides being the product used in the author's doctoral thesis, its use as a case study in our book is justified because: (a) it is a product required by more than 122 million persons in the world (Wheelchair Needs In The World, 2016); (b) it is expected to improve the users' quality of life; (c) it has a strong social appeal; and (c) it seems not to provide full user satisfaction.

Business in the wheelchair market seems to be very attractive. The *Fortune Business Insights* (2021) states that the global wheelchair market is expected to reach an estimated $8.09 billion by 2026 and is forecast to grow at a CAGR (compound annual growth rate) of 7% from 2018 to 2026. The size of such a market is sufficient to justify the use of mass-manufacturing techniques and marketing strategies in the production, sale, and distribution of wheelchairs.

It is true that designing, prescribing, and providing a wheelchair, or any other specific product for people with disabilities, is always a commitment. In the case of a wheelchair, it needs to be comfortable and safe, but also light and maneuverable. Many wheelchair users are not just dependent on them to move around. They also depend on relatives, friends, or carers to get them out or help them transfer and get into the wheelchair.

It is certain that designing, prescribing, and providing a wheelchair for a disabled person is always a compromise. The wheelchair needs to be comfortable and safe, yet also light and maneuverable. Many wheelchair users are not only dependent upon them for moving themselves about. They are also dependent upon relatives, friends, or a paid carer to take them out or to help them to transfer in and out of the wheelchair.

Thus, it is important to consider the needs of the carer (indirect user) as well as the needs of the wheelchair user (direct user).

This complex situation seems not to have been adequately addressed by wheelchair designers. This has resulted in recognizable design problems in several wheelchairs currently available on the private market. According to Cooper et al. (1997), the wheelchair has, for most of its history, been a design that has segregated instead of integrated its users.

It is currently accepted that usability is essential to guarantee the quality of products. Ergonomics plays an important role in guaranteeing usability and, consequently, better performance for consumer products in general and products for the disabled in particular.

QUESTIONS THIS BOOK AIMS TO ANSWER

A number of questions related to performance, quality, and the product design process need to be addressed such as:

- What is product quality?
- What can be considered an ergonomically well-designed product?
- What is the role of ergonomics and product design in product development?
- How does the product design process work?
- What is the role of users in the product design process?
- What is the role of ergonomics in the product design process?

- How do we improve a product's ergonomics and other specifications?
- What is the role of safety and standards?

This book assumes that meeting the users' needs is the key point to ensure quality in the product and consequently satisfying the consumer (quality is briefly discussed in Chapter 2). In view of this, some questions need to be answered such as:

- What can be defined as user needs and requirements?
- How do the needs of the disabled differ from those of the whole population?
- How have the needs of the disabled been met in the design of consumer products and products for independent living?
- What are the requirements involved in the design of products for the disabled and the able-bodied population?
- What are user satisfaction and dissatisfaction?
- How is user's satisfaction related to product design for the able-bodied population in general, and for the disabled in particular?

There is a need to investigate in more depth the design of products for the disabled population. This will hopefully lead to answers to the following questions:

- What determines the design of products which include the needs of the disabled?
- Is it possible to design products and devices so that they are usable by both the able-bodied and the disabled?
- What are the special characteristics of products for the disabled?

There are several methods in industrial design, engineering and manufacturing which are used to guarantee the competitiveness and acceptability of consumer products. It will be important to investigate the current methods in design and manufacture of products, which are based on user needs, to answer the questions below.

- What is product quality and how does it affect the manufacturing process?
- What is the role of user satisfaction in product development?
- What methods, based on user needs, are available in design and manufacture?

One of the main questions to be addressed in this book is if there is any involvement of users in the product design. The answer to this question will reveal if the "voice of the user" is being heard throughout the design process. Since the research that generated this book had a wheelchair as a case study, the first step in finding an answer to that question was to approach wheelchair designers. It involved answering the following questions:

- How do they approach the design of wheelchairs?
- How do they meet both physical and ergonomics specifications?
- What kind of data do they need from users?

The answer to these questions will be vital in revealing whether what is regarded as good practice in the design process is implemented by wheelchair designers and manufacturers.

Once again, it is important to draw attention to the fact that, although this book presents wheelchairs as case study and focus of analysis, the methodology presented here can be applied to any consumer product.

The result of the research generated the methodology that we will call the **Ergodesign Methodology for Product Design**. After produced the methodology, it was possible to ask some questions, whose answers served to validate it, and present a revised version of this methodology.

- How to evaluate the proposed methodology?
- How does a sample of designers evaluate the new methodology?
- What are the weaknesses and stronger points in the new methodology?
- Is the methodology acceptable, useful, and usable?

The answers to the above questions imply the adoption of a strategy to approach all the topics involved.

So, the main objective of this book is to introduce a user-centered product design methodology which can translate user needs into the design of products for the disabled and the able-bodied population.

To that end, the book is organized into the following chapters:

- Chapters 1 and 2 contain the introduction, some questions that this book intends to answer, analyze the design of ergonomic products, and analyze the relationship between ergonomics and design.
- Chapter 3 presents the product design process including the phases of design specification and some considerations about product safety, human error, and product usability.
- Chapter 4 analyzes consumer needs, satisfaction, and dissatisfaction.
- Chapter 5 analyze the design for the disabled and able-bodied people and some methods based on user needs for the design and manufacture of products.
- Chapter 6 introduces models and methods based on user needs for the design and quality of consumer products, Kansei and Kawaii engineerings, discusses the concepts and methods of usability and user experience, and reflects on the role of design and ergonomics in the pandemic period.
- Chapter 7 shows the steps involved in the human-centered product design methodology, investigates its suitability, and presents a step-by-step guide for using the methodology.
- Chapter 8 presents an overview of the book, its reflections and proposals, and discusses the methodology developed and its phases.
- The appendices include a summary of the methodology presented in graphic format and the results of the field research, conducted by the author, which gave rise to this methodology.

2 The Design of Ergonomic Products
Ergodesign

This chapter discusses the interaction of users with hundreds of consumer products in their lives. It is expected that this interaction will be done in a pleasurable, safe, and efficient way, free of physical damage, inconvenience, and accidents. We argue that this will only occur if the product has a good design, which presupposes that it is an ergonomically well-designed product. The concepts of consumer product, quality, ergonomics, ErgoDesign, and product design are presented. The sometime uneasy relationship between designers and ergonomists is also discussed.

2.1 ERGONOMICALLY WELL-DESIGNED PRODUCTS

Nowadays, consumer products, no matter how complex, are supposed to make work and leisure easier. In everyday life, users have to interact with hundreds of consumer products and they expect that these products will perform their activities in a quick, safe, efficient, and pleasant way. However, the many frustrations and errors that usually occur in handling a product show that this is not always the case. If this fact is true for consumer products in general, it also applies to those products used by the disabled population.

Cushman and Rosenberg (1991) and Wilson (1983) define **consumer products** as those used by the general public and distinguished from a commercial product by their characteristics, capacity, and speed of operation. Commercial products usually require trained and specialized operators, and this is not a requirement for consumer products, which are usually purchased by the users themselves. The authors state that consumer products fall into two categories: (a) those used for the satisfaction of more general human wants and needs and (b) those designed for specialized groups, such as children and disabled people. They are usually used in or around the home, in a residential or social setting rather than in a workplace environment. Consumer products often must interface with other products and systems (Roy, 2018). Users of these kinds of products are often untrained, unskilled, and unsupervised; may be any age, of either sex, or any physical condition; and may have widely varying educational, cultural, or economic backgrounds.

ATTENTION

Consumer products are defined as those goods and services which are used by the individual/general public which usually have an added value (Suryadi et al., 2018; Cushman and Rosenberg, 1991; Hunter, 1992; Kreifeldt, 2007).

DOI: 10.1201/9781003214793-2

Consumer products (e.g. television, microwave oven, smartphones) may change the habits and behavior of the society in which they are used. Consumers do not buy some consumer products just because of their inherent utility but also because of the subjective values attached to them including status. For example, the user who buys a Ferrari is not only interested in a vehicle that takes him from point A to point B but in the status that the brand represents.

Consumer products, including a number of products for the disabled population, are sometimes designed having in view just sales and profits. For this reason, a lot of products with poor design are regularly introduced in the market without taking into account the real consumer's needs. Thimbleby (1991), in *The Ergonomics Society Lecture*, at The Chartered Institute of Ergonomics and Human Factors, UK, said that "We are all faced with poor design" (p. 1269) and Norman (2013) concluded "Unfortunately, poor quality design predominates" (p. 41). Some of these products have a short life in the market due to consumer rejection, weakness before competitors, or litigation.

Stearn and Galer (1990) point out that it is at the consumer level that the effects of good and bad ergonomics, and consequently good and bad design, are most acutely felt.

The increase of competitiveness in modern consumer markets has stimulated companies to look for quality in their products and processes. Reducing losses during product manufacturing, reducing warranty claims, reducing product development cycle time, and improving user satisfaction are objectives of quality. According to Griffin and Hauser (1993), quality improvements lead to greater profitability. The concept of quality adopted in this book is user-based and is taken from De Feo (2017). He says that "quality consists of those product features which meet the needs of customers and thereby provide product satisfaction". The number of companies in the market of products for disabled people is a clue that competitiveness in this market is also growing. So, the necessity to deliver quality to customer, either able-bodied or disabled, is no longer optional, but a question of survival for companies. Indeed, quality is based on the customer.

Ergonomics plays an important role in guaranteeing usability and, consequently, better performance for consumer products in general, and products for the disabled in particular. While ergonomics has become a widely known and respected discipline, ergonomics attributes (such as ease of use, ease of learning, high productivity, comfort, safety, and adaptability) are largely used by the media as elements that will add quality to products they sell and be perceived by users as necessary for the fulfillment of their needs. About 20 years ago, Leonard and Digby (2003) stated that the appeal of an "ergonomic design" of the product seems to have merit in the eyes of the advertisers. We can see that, nowadays, the term "ergonomic" to define the quality of a product is widely used in several advertising campaigns, even if such products lack ergonomic quality.

Ergonomics is a discipline that has the human being as its principal focus. It is useful, in its practice, to collect data concerning the body's structure, functioning, behavior, and the environment where work is carried out. So, ergonomics uses data largely derived from the fields of anatomy, physiology, psychology, and engineering. Consequently, ergonomics also uses methods originally concerned with the acquisition and application of these data. Ergonomics looks at the whole system and is concerned with the context of use.

According to the IEA – International Ergonomics Association (2020), ergonomics can be understood as "the interactions between human beings and other elements of a system and the profession that applies theory, principles, data and methods to design, in order to optimize human well-being and the overall performance of the system". Note that this official definition of ergonomics uses the term "design" as a project, change, intervention. In this way, Ergonomics is involved in the analysis and design of products and systems of various levels of complexity in order to optimize the interface between the user and the product, the environment and system, be it a toy or an automobile, a control room at an international airport and the equipment, environments, and systems used by its operators.

NOTE

Product ergonomics is the area of study which aims systematically to analyze artifacts and their interaction with humans (Soares, 2012). According to the author, it deals with the use of ergonomic methods and tools for the analysis of objects manufactured on an industrial scale and prioritizes the relations of the user, operator, consumer, and those who maintain the product.

Ergonomics is a discipline supported by scientific data; product design is the process of creating new and improved products for the use of people and manufacturing aims to convert raw material into manufactured products, thus producing valuable and tradable goods. Ergonomics has clearly strong inputs from science while Product Design is assisted by aesthetic and subjective inputs. Manufacturers, on the other side, are mainly interested in the performance of the product in the market in terms of the quantity of goods sold and the profit made.

Usually, the three groups have different approaches. Ergonomists focus mainly on product usability and safety, employing empirical methodologies to achieve this purpose. Product designers endeavor to seek a balance between form, value, and appearance of products, relying on experience, intuition, and creativity to achieve this end. Manufacturers are more pragmatic having to fight for survival in an extremely competitive environment. For a long time that ergonomists have criticized designers for producing unsafe products, failing to emphasize the importance of usability and the lack of scientific reasoning (Grandjean, 1984; Wood, 1990). On the other hand, designers have said that ergonomic data are presented in a format or language unsuitable for designers. They represent an obstruction to design creativity.

The sometimes-uneasy relationship between designers and ergonomists has been mentioned by several authors (Pheasant and Haslegrave, 2018; Vincent et al., 2014; Abeni, 1988; Brown and Wier, 1982; Grandjean, 1984; Lingaard, 1989; Ryan, 1987; Smith, 2013; Ward, 1990, 1992; Wood, 1990). Although the previously cited authors recognize friction between ergonomics and product design, they are unanimous when they affirm that this disagreement needs to be overcome.

The successful integration of ergonomics and product design will produce an aesthetically pleasing and functionally superior product. They are both directed to the same end: fulfilling user satisfaction and producing a successful product. Harris (1990) claims that because the world markets comprise a multitude of anthropomorphic, behavioral and cultural differences, ergonomics knowledge is vital in helping design to meet the challenge of product development for a global market. So, the integration of ergonomics and product design seems to be particularly relevant when designing products which claim to be used for both the able-bodied and the disabled population.

REFLECT

One of the main areas of conflict between product designers and ergonomists arises from the emphasis that each group places on the methodology used to achieve its goals. Do you agree with this statement?

Designers are always expected to be innovators, always looking for a different solution to a problem, by the way they work in a creative and intuitive manner, trying out a number of solutions and evaluating them later. They usually approach the problems using what is called "lateral thinking", which means the use of creative thinking to solve problems avoiding a too logical and too constrained to conventional frames of reference approach. Ergonomists, although they sometimes use creative techniques, tend to use "vertical thinking", analyze the problem, and develop formulae or experiments that will deliver what they regard as the answer or best solution.

INTERNET

The use of Ergonomics in order to project an aesthetically pleasing product is a challenge of good design. Niels Diffrient, referred to by some as "the legend of design", was a pioneer in ergonomic design (Ergodesign) and in this video he details how he completely rethought the office chair studying the human body (in all its shapes and sizes).

Rethinking the way we sit down, Niels Diffrient, 2002.
https://www.ted.com/talks/niels_diffrient_rethinking_the_way_we_sit_down #t-76490

Ergonomics plays three traditional roles in product development: (1) user needs identification, (b) user interface design, and (c) test and evaluation. In fulfillment of these roles, ergonomists have appropriate procedures: (1) identifying user needs, (2) carry out the user interface design, and (3) test and evaluate the product. In fulfilling these roles, ergonomists have appropriate procedures and knowledge to: (a) identify user needs and preferences and verify how effectively these needs and preferences are met, and (b) measure how effectively user needs are met, in a way that allows them to provide this feedback at various stages of the product development cycle.

Ergonomics and, more specifically, product ergonomics, also called **Ergodesign**, can be considered a fundamental tool in the search for product design quality.

ATTENTION

Ergodesign is understood as the part of ergonomics that systematically aims to analyze artifacts and their interaction with humans. This term was adopted by the renowned physician and ergonomist Etienne Grandjean who organized Ergodesign 84, a conference focused on ergonomics and design for the electronic office (Grandjean, 1984). Moraes (2013) states that Ergodesign ensures the optimization of the development of ergonomic and design technologies in the creative process.

We adopted the term **Ergodesign** for our methodology because we understand that human-centered design is not possible without Ergonomics and **Ergodesign** represents exactly the relationship between ergonomics and design that we want to emphasize in this proposed methodology.

Ergonomically well-designed products (**Ergodesign**) are those which consider a wide variety of users – the everyday user, the curious, older people, children, male, female, the healthy, or unhealthy – offering safety, efficiency, comfort, and aesthetic satisfaction, under normal conditions of use, and under foreseeable conditions of misuse. Although, in general, not all user satisfaction factors are necessarily ergonomic, ergonomically well-designed products aim to guarantee user satisfaction.

Almost 15 years ago, Dirken (2007) stated that it is a sad truth in design and marketing that in most cases styling of products comes first, technology second, and ergonomics only third. It seems like things have not changed much in all these years. There is an unnecessary conflict between ergonomics and aesthetics (Andre and Segal, 1994). Norman (1988) argues that:

> If everyday design were ruled by aesthetics, life might be more pleasing to the eye but less comfortable; if ruled by usability, it might be more comfortable but uglier. If cost or ease of manufacture dominated, products might not be attractive, functional, or durable. Clearly, each consideration has its place. Trouble occurs when one dominates all the others.
>
> Norman (1988, p. 151)

The balance between these attributes will distinguish good and bad designs and, consequently, ergonomically well-designed products. Of course, this balance might be established based on the context created by the user, the task, the environment, and the culture. And also, in the case of products for the disabled independent living (called henceforth "products for independent living"), the user's medical and therapeutic needs. Designing implies a continuous choice between several solutions. The designer, for example, deals with conflicting interests between aesthetics and usability.

Norman (2013) has drawn attention to the fact that well-designed products, as opposed to poorly designed products, are easy to interpret and understand because they contain visible clues to their operation. The author refers to the principle *Form follows Function* that states products should indicate how and for what they were intended to be used. Certainly, this principle guides most product ergonomic designs. Meanwhile, it is important to observe that the introduction of new technologies – particularly with the use of electronics and microelectronics components – and therefore the possibility to produce miniaturized products, requires a new form of user interface – the medium of communication between the user and the product. The idea that form follows function is basic to industrial design, but this is not the only means that the design must fulfill the "given" function of the product: design is also a continuous interpretation of what function is about, of setting new functional demands and meeting them. This is quite evident with the inclusion of new forms of interfaces such as voice command, gesture interface and brain–computer interface.

INTERNET

Watch this delicious lecture by Don Norman "The three ways good design can make you happy". He talks with great humor about beauty, fun, pleasure, and emotion in design. Norman says that "Cognition is understanding the world, emotion is interpreting it". Norman lists three levels of design: visceral, behavioral and reflective.

3 ways good design makes you happy, Don Norman, 2003.
https://www.ted.com/talks/don_norman_3_ways_good_design_makes_you
_happy/transcript#t-1420

The concept of *Form follows Function* is largely used in most of disability products for independent living. The design of such products is generally initiated by the medical and therapy professions in response to a medical and physical need. Therefore, the design of such products is frequently guided to solve the problem within the context of the users' disability rather than design a product which takes into account the users' aspirations, desires, and lifestyle as well as fulfilling its functional role. According to Barber (1996), the result of this approach is that the design generally leads to a solution that is centered more towards a piece of technical apparatus than towards a product geared to meet the user's desires. This is absolutely true and easily observed in those products for independent living: designers often prioritize the medical and therapeutic requirements and forget the user needs in terms of their personal aspirations such as uniqueness, values, and status.

In looking for ergonomically well-designed products for both able-bodied and disabled people, ergonomics and product design perform distinct, but not incompatible, roles. Both fields of activities are responsible for defining the user interface. The role of ergonomics and product design will be discussed in the next sub-chapter.

NOTE

It is worth noting that when we use the terms "general population" or even "able-bodied or people without disabilities", we are dealing with relative terminology. This is explained why at some stage of life, as people who have suffered an accident and need support to walk or those who are considered elderly, for example, are faced with some type of disability, such as motor, visual, hearing, among others. So, it is a matter to be discussed whether someone can be considered a "person without a disability", since in various stages of life we may be "temporarily disabled".

2.2 ERGONOMICS AND PRODUCT DESIGN: BRIDGING THE GAP

Good design can be conceptualized as one that considers the **skills** and **limitations** of users in the development of the project, in order to make their **aesthetic**, **manufacturing**, and **usability** aspects compatible. Designing well is not easy. Product development is a risky business because it involves, with a high cost, many areas of the company. The design process can reduce the accidents and/or cost of product failure. To have a reasonable chance of success it should meet user requirements fully. The product design development process will be discussed in Chapter 3.

ATTENTION

From the point of view of ergonomics and usability, we define **Product Design** as a projectual technology that aims at product development, with a defined configuration, for manufacturing products in small or large series, considering issues of use, meaning, performance, operation, cost, production, commercialization, market, formal and aesthetic quality, and considering its environmental, urban, and ecological impacts.

After having conducted interviews with four designers, Mossel and Christiaans (1991) stated that aesthetic aspects are often so important for the designer that they can overrule the constructive, managerial, and ergonomic aspects. The study carried out by the authors led to the following conclusions: (a) most of the ergonomic information is taken from existing products, designers' presumptions, or by the clients themselves; (b) no users' trials are done; tests are carried out by the designers on themselves; and (c) ergonomic aspects have a low priority compared with aesthetic and managerial aspects. Apart from the conclusion stated in item (b), the remainder are similar to those found in the survey carried out by Soares (1990). Although this study only encompassed the work of four designers, the results must be considered as a source of reflection for the role of ergonomics in design activities. It seems that the findings by Mossel and Christiaans (1991) and Soares (1999) are still quite current today.

Pheasant and Haslegrave (2018) argue that designers need to move from the idea that ergonomics is a matter of applying data and develop a totally user-centered approach. A user-centered design is a method to develop products based on the needs and interests of the user, with an emphasis on making products usable and understandable.

INTERNET

In this video, Donald Norman explains what user-centered design is and how it works.

Principles of human-centered design, Don Norman, 10 August 2018. https://www.nngroup.com/videos/principles-human-centered-design-don-no rman/

Norman (2013) defines two fundamental psychological principles of design to make products understandable and usable: (a) providing a good **conceptual model** and (b) making things **visible**.

(A) CONCEPTUAL MODEL

It is important to observe that a **good conceptual model** allows us to predict the effects of our actions, that is, the reaction provoked in the product by our action. The conceptual representation of a system is defined from its mental model, that is, the user's interaction with the product or system is based on his or her previous experiences with other similar products and current observation on how to use the new product. This interaction has the power to provide a predictive and explanatory assessment for the user to understand the system and to perform a good interaction with the new product (Norman, 2013; Christiaans, 1989; Wilson and Rutherford, 1989). A good example is a user who bought a new high-end Blu-ray disc player. High-end devices are audio or video equipment with exquisite workmanship and top-quality materials manufactured for a special market niche. This user certainly must have had experiences with old cassette players, CD, and DVD players. The user must already have a mental model formed in the use of these products, acquired in the handling of older products, compared to the most recent ones, and will transfer this mental model to his new high-end device, thus contributing to a better interaction with the product. In other words, users bring back lessons from interactions (mental models) when they switch from one old product model to another new one.

People form mental models through experience, training, and instruction. According to Norman (2013), the conceptual model can be seen in three perspectives: (a) the **design model**, the designer's conceptual model; (b) the **user's model**, the mental model developed through interaction with the system, and (c) the **system image**, the visible part of the device, results from the product itself (including documentation, instructions, and labels). In operating unfamiliar consumer products, the user can have great difficulties in finding the appropriate way to handle the product and that these difficulties can be of a cognitive nature. This is because Mrs Cindy, in the fictional story in the introductory chapter, had so many difficulties in operating her new microwave oven. The designer expects the user's model to be identical to the design model. Problems arise when designers do not interact directly with users and assume that this premise is always true. Problems are also more serious when the products are used by disabled users, mainly those suffering from cognitive impairments.

NOTE

To understand a little more about mental models and their interaction with design, visit these two pages of the Interaction Design Foundation.

Mental models
https://www.interaction-design.org/literature/book/the-glossary-of-human-co
    mputer-interaction/mental-models
A Very Useful Work of Fiction – Mental Models in Design
https://www.interaction-design.org/literature/article/a-very-useful-work-of
    -fiction-mental-models-in-design

Designers are not typical users. On the contrary, they become so expert in using the object they have designed that they cannot believe that anyone else might have problems handling this product Norman (2013). Thimbleby (1991) stated, in a sarcastic way, that designers tend to design things for themselves to use and fool themselves that there is no problem with the design and suppose the fault is entirely the user's for not thinking.

(B) VISIBILITY

The second principle introduced by Norman (2013) is based on the **visibility** concept: the correct parts must be visible, and they must convey the correct message. The author says that when simple things need pictures, labels, or instructions, the design has failed. A typical consequence of reduced visibility is reduced feedback. Feedback is the response to the interaction provided by the product or system and acts as a way for users to adjust their mental model. In other words: feedback is the response of the product or system to the command exercised by the user.

On account of modern technology interacting has also been changed. In the past controls of several products were designed to be held, turned, pulled, and pushed, today they are designed to be merely touched, spoken, or commanded by gestual. In the near future, commands can occur through the mind, consequently, a new form of feedback has been produced: information once afforded by the movements of hands and fingers, the depression of buttons and switches, or the sound of clicks and cranks is either absent or has been replaced by gestures, voices, and mental commands.

Modern technology permits greater freedom for the designer to explore a product's aesthetical and formal aspects. On the other hand, the designers must pay more attention when reducing feedback. For example, a minimal tactile and kinesthetic feedback when pressing a key of a mobile phone, or a control of an electrical wheelchair or scooter, when driving, could result in the necessity of users looking simultaneously at the phone or the control and away from the road. In a study carried out by Vitorino (2017), it was concluded that the use of the gestural interface in the handling of software on the computer presented musculoskeletal problems to users, due to the maintenance of static postures for a prolonged time identified through digital infrared thermography (see more details in sub-chapter 6.4).

It is important to pay attention that usability is determined by the specific user, the specific kind of task, and the specific environment in which the interaction takes place. In this way usability is a variable that may change with time and activity. Aspects of usability and user experience will be discussed in sub-chapter 6.4.

The principles mentioned above make the users the focal point of the design. A so-called user-centered design approach claims to focus on users in all stages of product development. The need to focus on the customer and end user at all stages of development, obtaining relevant, meaningful, and applicable feedback; and accurate market research which facilitates forecasts of future customer requirements is a key to a successful product at the marketplace and objective of the proposed methodology. It should also be added that accurate market research, which facilitates the prediction of future customer requirements, is essential in the development of a user-centered or human-centered methodology.

The unique way to carry out a design process centered on the user is using ergonomics beginning early in the product development process. Such an approach has been supported by several authors (Sun et al., 2018; Santos and Soares, 2016; Soares, 2012; Ahram et al., 2011; Robert et al., 2012; Mital, 1995; Cushman and Rosenberg, 1991; Harris, 1990).

The use of rapid prototyping and usability testing has enabled ergonomics to provide input earlier and to work iteratively, making design problems easier to identify and design recommendations easier to support. The term "usability", here, is related both to obtaining user requirements prior to initiating the product design process and in the early stage of design, as well as to evaluating prototypes and products that have already been built. Rapid prototyping (or desktop manufacturing – DTM) is the producing of a three-dimensional prototype from a CAD (Computer-Aided Design) model. Human-centered approaches, together with the use of rapid prototyping techniques, can be used both in products aimed at able-bodied users and in the development of products for those who require assistive technology.

There are three tools that should be used over the product designer activity: (1) CAID (computer-aided industrial design), (2) Task Analysis, and (3) Usability Testing.

CAIDs (Computer Aided Industrial Design), derived from CAD (Computer-Aided Design), are software that help in creating the appearance of design and product development. CAID is software used for product design, as opposed to manual design, which allows an increase in efficiency in design changes, concept tests and general optimization. CAID software is more conceptual and artistic, CADs are more technical. The designs represented through CAIDs allow for rapid prototyping.

Examples of CAID are AliasStudio, ICEM Surf, NX Shape Studio, CATIA Shape Design and Styling, SolidThinking, and Rhinoceros.

An interesting point to be observed is that Task Analysis and Usability Testing have been in the domain of ergonomics for several years. These tools represent a new potential for designers and ergonomists to jointly create products that have the users as a fundamental part of the process rather than a recipient of it. Industrial design can now become a user-centered process based on user interactivity instead of on user adaptability. The product design process and a user-centered approach will be discussed in the next section.

TIP

Ergonomics and Design were the themes of three books by the author that can contribute to a more in-depth analysis on the subject.

Karwowski, W.; Soares, M.M.; Stanton, N. (2011a). Human Factors and Ergonomics in Consumer Product Design: Methods and Techniques. CRC Press. ISBN-13: 978-1420046281.

Karwowski, W.; Soares, M.M.; Stanton, N. (2011b). Human Factors and Ergonomics in Consumer Product Design: Uses and Applications. CRC Press. ISBN-13: 978-1420046243.

Soares, M.M.; Rebelo, F. (2017). Ergonomics in Design: Methods and Techniques. CRC Press. ISBN-13: 978-1498760706.

ISBN-13: 978-1420046281

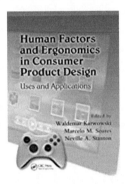
ISBN-13: 978-1420046243

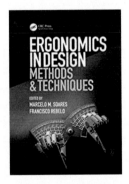
ISBN-13: 978-1498760706

2.3 SUMMARY OF THE CHAPTER

- Users of consumer products are often untrained, unskilled, and unsupervised. They may be of any age, of either sex, or any physical condition; and may have widely varying educational, cultural, and economic backgrounds.
- Users often do not buy consumer products just because of their inherent utility but also because of the subjective values attached to them.

- The increase of competitiveness in modern consumer markets has stimulated companies to emphasize quality.
- Reducing losses during product manufacturing, reducing warranty claims, reducing product development cycle time, and improving user satisfaction are objectives of quality.
- The necessity to deliver quality, as perceived by the user, to either the able-bodied or the disabled consumer, is a question of survival for companies.
- "Good design" can be conceived as one that considers the skills and limitations of users when developing a project in order to reconcile its aesthetic, manufacturing, and usability aspects.
- From the point of view of Ergonomics and Usability, Design is a project technology that aims at product development, with a defined configuration, for the manufacture of products in small or large series, considering issues of use, meaning, performance, operation, cost, production, commercialization, market, formal, and aesthetic quality; and considering its environmental, urban, and ecological impacts.
- According to IEA – International Ergonomics Association (2020), Ergonomics can be understood as "the interactions between human beings and other elements of a system and the profession that applies theory, principles, data, and methods to design, in order to optimize human well-being and the overall performance of the system".
- Ergonomics plays an important role in guaranteeing usability and, consequently, better performance for consumer products in general, and products for the able-bodied, and the disabled population, in particular.
- The balance between aesthetics, usability, and manufacture/technology will distinguish good and bad designs and, consequently, ergonomically well-designed products.
- Ergonomically well-designed products are those which consider a wide variety of users – the everyday user, the curious, old people, children, male, female, the healthy, or unhealthy – offering safety, efficiency, comfort, and aesthetic satisfaction, under normal conditions of use, and under foreseeable conditions of misuse. Although, in general, not all user satisfaction factors are necessarily ergonomic, ergonomically well-designed products aim to guarantee user satisfaction.
- The design of products for the disabled is frequently guided to solve the problem within the context of the users' disability rather than taking into account the users' aspirations, desires, and lifestyle as well as fulfilling its functional role.
- Ergonomics and Product Design perform distinct, but not incompatible, roles in the design of ergonomically well-designed products for both able-bodied and disabled people.
- Although recognizable friction between ergonomics and product design has been mentioned by a number of authors, they are unanimous when they affirm that this disagreement needs to be overcome.

- Ergonomics plays three traditional roles in product development: (1) identification of user needs, (2) user interface design, and (3) test and evaluation.
- There are two fundamental psychological principles of design to make products understandable and usable: (a) providing a good conceptual model and (b) making things visible. These principles make the users the focal point of the design.
- The unique way to carry out design, with the human at the center, is to use ergonomics from the beginning of the product development process.
- There are three main techniques which are currently used by product designers: (1) CAID (Computer-Aided Industrial Design), (2) Task Analysis, and (3) Usability Testing. These latter two have been used in the domain of ergonomics for several decades.

3 The Product Design Process

The concepts of Product Development and Product Design Process are presented. The chapter discusses what the product should do to meet the user's needs and presents a wide range of bibliographic references in several important areas for product design. The various stages of the consumer product design process (Design Specification, Concept, Modeling and Prototyping, Product Evaluation, Production and Marketing, and Further Evaluation) are presented and analyzed. Special attention is given to product safety which includes an analysis of human error and responsibilities in product design and manufacturing. Attention must be given to product liability which usually results from the application of negligence, breach of warranty, or strict liability in tort.

3.1 PRODUCT DEVELOPMENT AND THE PRODUCT DESIGN PROCESS

According to Kotler and Armstrong (2018):

> A product is anything that can be offered to a market for attention, acquisition, use or consumption that might satisfy a want or need. It includes physical objects, services, persons, places, organizations, and ideas.

Additionally, **Product Development** (which comprises the product design and engineering process) is the set of activities beginning with the perception of a market opportunity and ending in the production, sale, and delivery of a product (Ulrich and Eppinger, 2019). This concept applies to products aimed at the non-disabled and disabled user. Smith (2013) ponders that although products or services satisfy a want or a need, some market segments – car manufacturers, for example – do not sell the product itself; they sell the skills of their workforce, self-esteem, comfort, safety, and style. The product is part of the medium through which these needs and wants are satisfied. So, concludes the author, the user is essential to help manufacturers answer the question, "what do you provide?".

Marketing strategies are used extensively to supply products and services that appeal to varying types of people based on an understanding of people's aspirations, uniqueness, values, and status. However, these strategies have not been fully applied to products for independent living, largely because the main consumer is general health care and government agencies and not the people who use them.

This book assumes that product requirements (also called "product specifications") mean the precise description of what the product has to do. **User needs** are generally expressed in the "language of the user", that is, the way the user verbalizes his wishes and needs to be met by the product using his own words. Product requirements must specify, in an unambiguous, precise, and measurable way, what the product has to do in order to satisfy user needs (Ulrich and Eppinger, 2019).

The **product design process** is a series of compromises between several product requirements: function, performance, reliability, usability, appearance, and cost. Finding the exact design solution is sometimes very difficult, and a compromise has to be established between some acceptable solutions. Product design is the first and perhaps the most important step in the manufacturing sequence. The level of complexity in the product development process varies according to the nature of the product.

Cushman and Rosenberg (1991) define **consumer products** as those used by the general public and, according to the authors, differ from commercial products due to their characteristics, capacity, and speed of operation.

ATTENTION

The literature presents a wide range of references related to specific topics in product development. Below are some specific areas and their references.

- The **product design process** has been studied by many researchers, such as Gomes Filho (2020), Olsen (2015), Cuffaro et al. (2013), Milton and Rodgers (2013), Hanington and Martin (2012), Löbach (2001), Cross (2008), Baxter (1995), Jones (1992), Maldonado (1977), and Rozenburg and Eekels (1995).
- The **role of ergonomics and users** in the product design process was studied by Privitera (2019), Soares and Rebelo (2017), Iida and Guimarães (2016), Ulrich and Eppinger (2019), Wendel (2014), Karwowski et al. (2011a, b), Kreifeldt (2007), Cushman and Rosenberg (1991), Mital and Morse (1992), and Wood (1990).
- **User-centered design** is analyzed by Govella (2019), Endsley (2017), Still and Crane (2016), IDEO (2015), and Goodwin (2009).
- **Usability** and **user experience** are the focus of Whalen (2019), Pannafino and McNeil (2017), Mash (2016), and Jordan (1998a).
- **Design thinking** is the object of study for Curedale (2019), Pressman (2018); Luchs et al. (2015), McKay (2013), Albert and Tullis (2013), LUMA Institute (2012), Reiss (2012), Goodman et al. (2012), and Dumas and Lorign (2008).
- Aspects related to **quality** and the **production process** were studied by Anderson (2014), Magrab (2009), Cross (2008), Juran (1992), Mital and Anand (1992), and Pugh (1991).
- Some authors have focused on **universal design** and **product design for independent living**, including Vanderheiden and Vanderheiden (2019),

Hamraie (2017), Vanderheiden and Jordan (2012), Torrens (2011), Pullin (2011), Lidwell et al. (2010), Kumar (2007, 2009), Clarkson et al. (2003), Poulson et al. (1996), and Wilkoff and Abed (1994).

- The **emotional and pleasurable design** is the subject of the following references: Pavliscak (2018), Chapman (2015), Norman (2008, 2013), Walter (2011), and Jordan (2002).

The **product design process** can be defined as a method composed of a set of rational and systematic procedures with the objective of conceiving and developing physical products to be employed by users. Although extremely useful, the product design process itself is not sufficient to guarantee the good quality of the design of any particular product. In fact, no one can anticipate all of the problems that will arise during the design process, but the risks as well as the costs can be minimized by following good practice and using effective methods and appropriate information wisely (Poulson et al., 1996).

The design process for consumer products can be summarized as consisting of six main sequential, or sometimes concurrent, phases: (1) Design Specification; (2) Conceptualization; (3) Modeling and Prototyping; (4) Product Evaluation; (5) Production; and (6) Marketing and Evaluation. These six steps are now described.

3.2   DESIGN SPECIFICATION

The establishment of the broadest conceptual objectives that a new product will fulfill is defined in this step of the design process. Aspects such as defining the business plan for the new product, defining which needs the product will satisfy, who will use it, and what are its characteristics should be carefully analyzed. This phase is traditionally defined by marketing and management teams or from a discussion between designers and clients. Ergonomics might perform an essential role in the several steps which comprise this phase. It is important to say that design specifications are only one part of the total list of specifications in a product development process. Other specifications may include marketing, engineering, manufacturing, financial, and so on.

ATTENTION

The **Design Specification** is the phase responsible for:

- Identifying user needs;
- Evaluating competitive products;
- Establishing user profile;
- Defining product performance requirements; and
- Determining design constraints.

3.3   GENERATION OF DESIGN CONCEPTS

This phase of **Generation of Design Concepts**, also known as **Conceptualization**, involves the generation of ideas that fulfill the criteria previously established in the design specification. This process is usually based on designer creativity and intuition and how others have resolved similar problems. Various well-developed techniques such as brainstorming, synectics, and others are available for such purposes. An early objective of this phase is to produce as many solutions as possible without criticisms. From an ergonomics point-of-view, the problem with this approach is that the solutions engendered are rarely evaluated on the basis of safety or usability, resulting in the manufacture of many unsafe or inconvenient products.

An evaluation and selection of the best ideas are carried out. The use of a decision matrix, including the product specifications, helps in the choice of the best concepts (see sub-chapter 7.2.5).

ATTENTION

Ergonomics may contribute to this process of generating concepts providing the designers with an understanding of the users' physical and cognitive needs in order to generate solutions sensitive to function.

The concepts produced at this point are represented in the form of renderings and drawings detailed enough to form a clear idea of what the final product will be. It is important to mention three techniques used to design products: computer-aided industrial design (CAID), *Kansei Engineering*, and augmented and virtual reality. *Kansei Engineering* is a technique developed essentially to interact with users and his or her emotion. It is described in more detail in sub-chapter 6.3.

Computer-aided industrial design is a computer-based design system that allows the designer to create and evaluate product designs in three dimensions (3D) and to generate photorealistic images and animation from the basic geometric design (Chang, 2016; Onwubolu, 2013; Erhorn and Stark, 1994). As previously mentioned, there are several software systems available for using CAID. There are even options that connect to rapid prototyping printers, mechanical and laser cutting machines or generate files that can be read by them. The physical prototype generated by this technique can be used for tests and evaluations by users. The use of the product can be simulated from virtual or augmented reality.

Virtual and augmented reality technology helps designers to produce virtual models derived from the design's concepts, providing a greater user interaction. Wang (2002) points out that a "Virtual prototype, or digital mock-up, is a computer simulation of a physical product that can be presented, analyzed, and tested from concerned product life-cycle aspects such as design/engineering, manufacturing, service, and recycling as if on a real physical model". Rebelo et al. (2011) introduced the methods and applications of virtual reality in consumer product design.

INTERNET

Virtual and augmented reality are technologies that are increasingly present in our world and also in the activity of design. Do you want to understand how these technologies works?

(1) See the digital newsletter from Digital Arts: "Everything you need to know about designing with augmented reality"
https://www.digitalartsonline.co.uk/features/hacking-maker/everything-you-need-know-about-designing-with-augmented-reality/
(2) See the website "How Stuff Works?: How virtual reality works".
https://electronics.howstuffworks.com/gadgets/other-gadgets/virtual-reality.htm

Figure 3.1 shows the concept of a smart toy for children's music-learning. The product was developed, under my supervision, by the design student Hong-Chin Tu from the School of Design, Hunan University, China, in 2019. The product simulates the use of a violin, a piano, and a ukulele. The figure illustrates the different phases of the design project.

3.4   MODELING AND PROTOTYPING

Modeling is the phase of the design process responsible for the selection and development of the most promising concepts and turning them into representative static models (computer graphics or non-working "mock-ups") and working models. The objective of this phase is to produce realistic models suitable to meet the specifications and goals set out for them. Foam or other materials are used to create basic forms or mock-ups. It is important to observe that non-working models and "mock-ups" can also be used in the previous phase to help in the choice of the best concept(s). Models can then be transformed into working, full-scale prototypes.

Rapid prototype technology permits the production of a prototype in a few hours compared to the days or weeks of conventional prototyping, decreasing the cost and time required to create a physical model of a design. Figure 3.2 shows the conceptual and functional 3D models of a smart toy for children's music-learning.

Turning a concept into a physical reality involves the use of several numeric values: e.g., lengths, weights, diameters, and balance. At this point, ergonomics provides helpful support to the design in the form of extensive data from its literature. Data regarding posture, dimensions, and strength applied to the design and use of objects are provided by anthropometrics and biomechanics.

Anthropometry is a discipline based on Physical Anthropology, which studies the dimensions of the body segments of the human being. Anthropometric data define the measurements of size, weight, and structure of the human body applicable to the correct dimensioning applied to the design of products, equipment, and workstations.

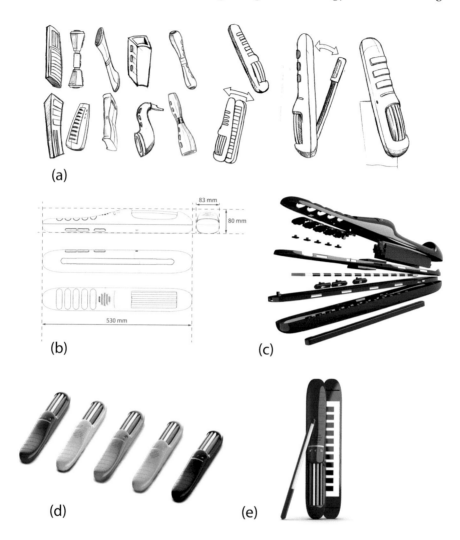

FIGURE 3.1 Concept generation of a smart toy for children's music-learning. (a) Defining design alternatives, (b) Dimension, (c) Explosion view, (d) Choosing color, (e) CAID-generated final design concept. Images courtesy of the author Hung-Chin Tu, School of Design, Hunan University, China.

According to Iida and Guimarães (2016) and Roebuck et al. (1995), anthropometry can be **static** or **dynamic**. **Static anthropometry** comprises the physical dimensions of the human body at rest or with few movements. Static anthropometry deals with the structural dimensions of the body, taken with the subjects in fixed and standardized positions: heights, widths, lengths, and perimeters. In turn, **dynamic anthropometry** measures the range of movement. Although the movements of each part of the body are measured, keeping the rest of the body static, it is observed that, in practice, each part of the body does not move in isolation, but there is a combination of several movements to perform a function. For this, it is necessary to define

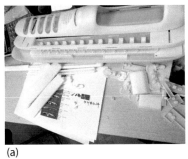

(a)   (b)

FIGURE 3.2   Prototype of a smart toy for children to learn music. The image is comprised of two figures. In the first, the prototype under construction (3.2a), in the second, the finished prototype, the student playing the toy in his violin version (3.2b). Image courtesy of the author Hung-Chin Tu, School of Design, Hunan University, China.

the movement angles to correctly size the use of a product or workstation. Both static and dynamic anthropometry provide data that should be used for the design of the first design models. More information on anthropometry can be obtained from Pheasant and Haslegrave (2018), Panero and Zelnik (2016), and Tilley and Dreyfuss Associates (2001).

INTERNET

Watch the video on the importance of designing considering extreme users (the biggest and the smallest). In this video, Sinéad Burke, 105 centimeters (or 3'5") tall, presents the difficulties she faces for a world designed for taller users. She wonders what it is like to design for a world considering small people and asks: Who are we not designing for?

Why design should include everyone, Sinéad Burke, March 2017.
https://www.ted.com/talks/sinead_burke_why_design_should_include_every
    one

Biomechanics is the "application of mechanical principles, such as reach and strength, to the analysis of the structure and movement of parts of the body" (Chaffin et al., 2006). Biomechanics provides, among others, data on mass, center of gravity, and the moment of inertia of the product that must be considered both for the body segments and for the objects that are lifted and/or driven.

Incorporating the users and hearing their voices, beginning with the very earliest product development process steps, contributes to reducing to a minimum users' resistance to the final design and the need for substantial modifications. In the case of products for people with disabilities, it can contribute to the acceptance and reduction of possible user resistance to the final design and the need for substantial modifications. The next sub-section analyzes in detail aspects related to product evaluation.

It is important to draw attention to the importance of considering users from the initial stages of the product development process. This helps to identify users' needs through their own voice. In the case of products for people with disabilities, it can contribute to the acceptance and reduction of eventual resistance from users to the final design, in addition to the need for substantial modifications. The next sub-chapter looks in detail at the aspects related to product evaluation.

3.5   PRODUCT EVALUATION

Product evaluation can be used from the design phase of product development with the evaluation of the first *mock-ups* until the evaluation of advanced prototypes at field sites and the "final prototype" (the master copy of the product which will later be mass-produced). Analyzing the interaction between product performance and users (either the disabled or able-bodied) from the results of product tests may indicate that some modification to the design is necessary. The tests also provide manufacturers with knowledge of the degree to which a prototype fulfills market needs and legal requirements.

The measurement of the interaction between consumers and products can provide requirements to improve the products' ergonomic specifications and general qualities. Only by such an approach can inadequate designs be identified. Specific methods are used for this purpose, including evaluating design in terms of factors such as safety, effectiveness, robustness, reliability, comfort, dimensional compatibility, ease of use, aesthetics, and, increasingly, pleasure aspects.

Broadly speaking, there are two kinds of product tests: **physical** and **ergonomic** (Cushman and Rosenberg, 1991). **Physical tests** are used to verify the product's technical quality, such as its physical, electrical, and electronic characteristics (e.g., potency, power consumption, impact resistance, corrosion, and resistance). Physical tests are relatively more important for the components, the product's internal parts, or products that usually have few contacts with users. The human does not interact in a major way with the product (or its components) in this kind of test.

The use of **ergonomic tests** on products is different from physical testing because the former involves the user directly and relates his or her anatomy, physiology, and psychology to various features of the product. These kinds of tests typically apply to consumer products for able-bodied and/or disabled people where the products are used in the home, work, and in leisure, where a major feature of product use involves extensive contact with people.

In this methodology, **Usability tests** are part of the ergonomic tests. Usability tests are an important and essential part of the product development process, mainly when intuition is used to make design decisions. They are concerned both with obtaining user requirements prior to or initiating the product design process and in the early stages of design; and with evaluating products that have been built. This approach makes users the focal point of the design. This topic will be discussed in more detail in sub-chapter 6.4.

The necessity to evaluate consumer products physically and ergonomically comes from the necessity of manufacturers to evaluate their new products and compare them to those already on the market, especially those of competitors.

Evaluating the ergonomic qualities of a product should be assessed at any phase in the design process. Computer-aided systems constitute excellent tools to carry out this early evaluation. Simulations with mock-ups, models, and prototypes permit the study of what the users' reactions will be during real-product performance and identify failure or malfunctions in the product.

The use of product evaluation contributes: (a) to keeping a company's good image (avoiding the danger of negative oral propaganda); (b) to avoiding negative reactions from consumer organizations; and (c) to avoiding liability in court.

The criteria used to evaluate products for independent living are not very different from those to evaluate consumer products in general and include the following aspects:

- **Safety**, the property of being able to handle a product without the risk of damage, death or injury provoked by faults, malfunctions or errors in normal use, or foreseeable misuse, of the product or its components
- **Effectiveness**, the characteristic of a product which enables it to do the job it is intended to do efficiently and effectively with a reasonable amount of human exertion to produce the intended effect
- **Suitability**, the characteristic a product appropriate to the user's social and therapeutic requirements (in the case of products for the disabled person)
- **Robustness**, the quality of a product to be able to resist fairly hard use and occasional misuse
- **Reliability**, the probability that an item will perform a required function under conditions for a stated period of time; it means having confidence that the product will perform its intended function satisfactorily for a defined period of time.
- **Comfort**, the quality of a product to produce physical and mental well-being during any activity associated with its use
- **Dimensional compatibility**, the characteristic of a product to be dimensionally suitable with the anatomical and anthropometric dimensions of users and the physical constraints of the environment in which it is to be used
- **Ease of use**, the attribute of a product not to demand excessive strength, over-exertion, or attention in use
- **Aesthetics**, the virtue of a product to be pleasurable to the user in terms of its visual appearance, sound, smell, and feel
- **Good value**, the ability a product has to offer good value for money at purchase, in maintenance, and in the repairing of parts and components

It is important to mention the need to be flexible for some product features such as labeling and controls to address accessibility problems that are identified.

The criteria for evaluating products for able-bodied or disabled people have to be specified in detail to be applicable to the testing of particular aids. This specification should include the following:

- For the **product**: dimensions, materials, components, controls, displays, instructions, structure, noise, vibration, and any special product features.

- For the **user**: age, sex, anthropometry, senses, intelligence, functional ability, socio-economic status, product ownership, and any special features of the users.
- For the **task**: the objective to be achieved by the use of the product and the dynamic interaction of the user, the environment, and the task for which the product was designed.

3.6 PRODUCTION

The **Production** phase involves a variety of activities, including process and material selection, production operation planning, material handling, inspection, and quality control and packaging. Manufacturing, the essence of the production phase, has the goal to accomplish the conversion of raw materials into finished products as easily, quickly, and economically as possible. It requires that the following steps be taken: product design, manufacturing system design, and manufacturing system operation.

According to Vanlandewijck et al. (2019), an extremely difficult problem is faced by the manufacturers of assistive technology. On the one hand, the manufacturers need to produce the highest possible volume of products to reduce manufacturing costs. On the other, these products should suit the user's individual capabilities and limitations. The use of the universal design approach (discussed in sub-chapter 5.3) and modular design, which permits the combination of a number of variants for each component of the product in such a way to meet individual needs, may be the solution to this problem. However, the more specific the product is in meeting individual needs, the more difficult it is to use the above-mentioned approaches. A disabled person with very specific needs, like the late physicist Stephen Hawking, for example, could not be accommodated by the universal design or modular design approach.

Industrial engineers are responsible for solving production problems. This phase of the product development process does not involve directly product ergonomics and industrial design and is not discussed in depth in this book.

3.7 MARKETING AND FURTHER EVALUATION

Marketing and evaluation of the product are responsible for assessing customer feedback before and after the product has already been launched onto the market. Using appropriate techniques, the marketing team is able to define strategies to identify customer satisfaction/dissatisfaction after the purchase. Obtaining feedback will determine the image and performance of the product and allow immediate action to be taken if any problems arise.

Once introduced in the market, the new product normally goes through one or more periods of sale growth and decline. This phenomenon is called the Product Life Cycle. According to the Product Life Cycle Stages, the life cycle of a product comprises five stages: Introduction, Growth, Maturity, Saturation, and Decline (Product Life Cycle Stages, 2019). Poor design and lack of good ergonomics can severely affect the product life cycle. Wilson (1983) claims that, from an ergonomic point-of-view, if all stages in this cycle are carefully considered, some requirements may be produced incorporating ergonomics design criteria (ergodesign). This procedure will limit constraints in user interaction, improve safety, and facilitate recycling. In this way, it will contribute to extending product life.

Although designers and ergonomists should consider the product life cycle, which includes product sales and disposable phases, the phase of **Marketing and further evaluation** is not directly involved with design (but it is very useful for the next generation of designs) and, in view of this, is not discussed in this book.

3.8 PRODUCT SAFETY

An important point that should be addressed in the product design process is related to product safety. This matter affects both the disabled and non-disabled users.

Product safety is an extremely important item in the design process because consumer products frequently harm their users. Reasons for this are many and varied and include misuse, faulty manufacture, and even poor design. Consumer products that do not attain safety requirements may cause injury or death to users and be excluded from the market by preventive or repressive legislation. Consequently, financial loss and negative publicity can produce catastrophic effects for the company.

Laughery (1993) states that products are frequently designed requiring some knowledge or information on the part of users which they – at least some of them – may not or do not have. This doctrine assumes that consumers will use their intelligence and experience to protect themselves against possible hazards while handling products.

Designing a consumer product based on a safety approach is an activity that must consider the interrelation between the product itself, the user, and the environment presenting a low risk to the user. Normal use and predictable misuse should also be taken into account, especially when the use of the product involves children, the elderly, and people with disabilities. It is extremely important that industrial design and manufacturers are fully aware of the potential for accidents associated with the product they produce.

Details of home and leisure accidents serious enough to warrant a visit to an accident and emergency department were collected from 1978 to 2002 at a sample of 16 to 18 hospitals across the United Kingdom (ROSPA, 2020). These data served to provide an in-depth understanding of how and why home and leisure accidents happened so that interventions could be implemented, as well as strategies to prevent them from happening again in the future.

NOTE

Data from ROSPA (2020) points out that, in the United Kingdom, accidents in the home result in over 6000 deaths per year. The data obtained during all the years of the survey revealed that more than two million children under the age of 15 experience accidents in and around the home every year, for which they are taken to accident and emergency units. The cost to society of U.K. home accident injuries has been estimated at £45.63 billion during the research period.

According to the United States National Safety Council, in 2017, an estimated 127,300 preventable injury-related deaths occurred in homes and communities, or about 75% of all preventable injury-related deaths that year in the United States (Injuryfacts, 2019). The Council points out that over the last ten years, home and community deaths have increased by 60%, and the death rate per 100,000 people has increased by 49%. This involves a cost of US$ 472.6 billion. These data give the clear impression that the home is the most dangerous single location in Great Britain and the United States, more dangerous than the road because the data on domestic accidents are superior to traffic accidents. Consumer products, as a whole, make a considerable contribution to this statistic.

The statistical data at several websites shows that the high incidence of accidents in the home is predominant among the very young and old people. A possible cause of this is because these people spend more time at home. Furthermore, children are inquisitive and unaware of the dangers that surround them; older people have their mental and physical skills decreased, and so are more vulnerable. This physiological reduction in physical and mental abilities is due to the normal aging process.

Factors that lead to accidents are predominantly present during product use and are dependent upon: (a) the product design, (b) the environment in which the product is being used, and (c) the characteristics and behavior of the user. Ergonomics can provide a major contribution to the field of product safety, ensuring that the user is fully considered at the several product development stages.

INTERNET

The New Zealand Government's Commerce Commission published a nice video on 5 December 2018, alerting manufacturers, exporters, and traders to criminal liability for accidents related to the products they manufacture, export, and sell. The video warns of the need for safety testing, particularly with toys for children under three years of age.

The story of a toy – Guidance for businesses, Commerce Commission, New Zealand, 2018.
https://www.youtube.com/watch?v=FRW8wh2k7nY

Kreifeldt and Alpert (1985) describe the following steps as necessary for the design of safe products:

- Conduct a **potential hazard analysis** using members of the representative user population in order to investigate the use and foreseeable misuse of the product;
- Determine the **characteristics of the user population** for each foreseeable use or misuse of the product;

- Identify ways in which **physical risk** can occur, for each use and for each relevant part of the user population;
- Make a **comparative list of competing products**, focusing on potential risks and hazards, and make sure that your product will perform better;
- Check the **state-of-the-art** through appropriate literature;
- Be sure that your product is suitable for all possible and relevant **rules and regulations**;
- Analyze the intended **use environment**;
- Conduct **tests** that can indicate which potential hazards may arise;
- Provide, where appropriate, **warnings,** and **user manual**.

Kreifeldt (2007), Christensen (1987), and Ryan (1985) suggest the following strategy for reducing risks in the use of the product, to be carried out in the stages of product development:

- **Eliminate the potential hazard** in the design stages and
- **Reduce the potential hazard** by providing adequate protection or security features that reduce the risk.

In those cases where the potential hazard cannot be eliminated, the designer must:

- Provide **physical barriers** to the user, preventing contact with sources of potential danger;
- Provide, in the product, appropriate **legends and danger signs**;
- Provide adequate **instructions** on the safe use of the product; and
- Provide necessary **training** on the safe use of the product.

To achieve this goal, Ryan (1987) states that ergonomists and designers have many valuable procedures and diverse resources available, such as:

- Organizing a **Product Safety Program**.
- Employing **product safety assessment techniques**.
- Using **ergonomic principles** to: (a) Identify users (e.g., age, skill, strength, memory, and fatigue) and (b) Interface analysis (e.g., expected use and unexpected use).

The use of mock-ups and models may be necessary to test anthropometric and bio-mechanical adequacy. In addition to the product itself, protections, instructions, and warnings must be analyzed to identify incorrect or inappropriate elements, use of inappropriate codes or standards (which are not in accordance with the state-of-the-art), and the use of inappropriate materials.

Acioly (2016) conducted a study with safety instructions on food product packaging and concluded that a system using safety instructions in augmented reality showed better results than in the physical system.

TIP

The contribution of design is fundamental to the production of safe products. A number of authors highlight the importance of design to improve the safety of products, such as Gullo and Dixon (2018), Pheasant and Haslegrave (2018), Li and Lau (2018), Abbott and Tyler (2017), Sadeghi et al. (2017), Zhu et al. (2016), Cushman and Rosenberg (1991), and Jenkins and Davies (1989).

In terms of product safety for people with disabilities, Copper et al. (1997) state that some wheelchair accidents occur as a result of poor design.

A product can be defective in two ways: (a) products which were not produced as planned but which include some manufacturing fault, or which were incorrectly inspected and (b) products which were produced as planned, but which are dangerous to the public or to their owners. In fact, it is not sufficient just to design products that are safe when used as intended; improper use must also be considered.

Falling and tipping-related accidents are the primary accidents connected to wheelchair use. In 2003, more than 100,000 wheelchair-related injuries were treated in emergency departments in the U.S., double the number reported in 1991 (Xiang et al., 2006). The authors state that wheelchair-related injuries may have increased in the U.S. during the past decade. Prevention efforts should address the interacting complex factors that influence the risk of injury while using a wheelchair.

The product's failure frequently occurs shortly after the consumer had purchased the product. During the mid-life stage, consumers can expect, on most products, a relatively long period of reliable and safe use. Failures in this stage may be attributable to unforeseen changes in product use. In the later stage of product life, when it begins to wear out, failure probability increases.

At this point, product failures are usually caused by accumulated stress in materials, abrasion, environmental factors, etc. Some physical tests can be used to test materials and components to predict product failures. Ergonomics is likely to be most useful in foreseeing faults during the initial stages of the product design process.

Cleverism (2019) states seven reasons why most products fail after being launched into the market:

1) Failure to pinpoint and understand consumer needs and demands;
2) Wrong price to the launching of a new product into the market;
3) Product attend a non-existent problem;
4) Targeting the wrong market;
5) Poor preparation to launch the product into the market;
6) Technological inadequacies; and
7) A launch of a revolutionary product that the user does not understand how to use it.

Additionally, the author recommends talking with the target user to understand their needs.

3.8.1 HUMAN ERROR

Reason (1990) defines human error as something was done in a way that was not intended by the actor; was not desired by a set of rules or an outside observer; or that led the task or system outside its acceptable limits.

Until quite recently, product failure was essentially attributed to user error. Although this is no longer the main approach, human error continues to be an important and useful point to be considered during a hazard analysis. In this context, a hazard can be understood as a set of circumstances (conditions/situations) that have been associated with their potential risk of causing injury (Consumer Reports, 2019; Cushman and Rosenberg, 1991; Stadler-Estrin and Estrin, 1987).

The Canadian Centre for Occupational Health and Safety (2019) and the website of the HSSE World – Health, Safety, Security and Environment (HSSE World 2021) point out the difference between **hazard** and **risk**. According to these sources, a **hazard** is a potential source of harm to a worker or user; **harm** is a physical injury or damage to health; the **risk** is the combination of the likelihood of the occurrence of a harm and the severity of that harm.

Cushman and Rosenberg (1991) state that the potential danger associated with products can be classified as **obvious** and **latent**. According to the authors, the **obvious potential dangers** are inherent in some products, for example, weapons and fireworks. On the other hand, the **latent dangers** are those associated with the potential that a product has to cause a dangerous situation, for example, products that have moving parts (such as lawnmowers, tools for pruning fences, and food processors). Still, according to the cited authors, users cannot be expected to recognize and adequatcly assess the potential latent danger.

Human error can occur wherever human beings are involved in carrying out tasks. With users handling consumer products, it is not different. However, human error can be controlled and sometimes predicted before injury or damage occurs. The startling increase in the complexity of products will inevitably lead to errors and problems when in use.

When people interact with a product, they will often be engaging in some form of a problem-solving exercise. The best solution to perform the action is dictated by such factors as available information, state of the product, user's cognitive repertoire, and the user's experience with other similar products (Baber and Stanton, 2002). Task analysis can be considered a valuable tool for error identification.

TIP

Want to know more about human error? Read: Dekker (2014), Woods et al. (2010), Peters and Peters (2006), Baber and Stanton (2002), Kirwan (1992a, b), and Reason (1990).

Casey (1998) introduces 20 factual and arresting stories about people and their attempts to use modern technological creations. The author shows how technological failures result from the incompatibilities between the way products are designed and the way people actually perceive, think, and act.

3.8.2 PRODUCT-SAFETY ANALYSIS, STANDARDS, AND REGULATIONS

The degree of hazard associated with a product is often difficult to quantify, but there is a point when the hazard, under specific circumstances, will increase the risk to the degree that the probability of injury is great enough to make it predictable. In order to guarantee that products should not contain or present hazards that may cause injury to the user or persons coming into contact with the product, safety analysis, and tests should be carried out.

REFLECT

Have you ever been or know someone who has been the victim of an accident involving consumer products? If so, who do you think was to blame? From user? From the designer? From the manufacturer?

The design of consumer products, and also products for the disabled, is subject to many governmental standards, regulations, local codes, and product standards with an emphasis on safety. As most of them have the force of law, the first step in the initial design stages is to verify what law, codes, standards, and regulations are applicable to the design problem.

ATTENTION

The Consumer Product Safety Commission from the United States Government has stated the "Consumer Product Safety Improvement Act (CPSIA)". It is a document with significant new regulatory and enforcement tools as part of amending and enhancing several CPSC statutes, including the Consumer Product Safety Act.

The document can be a guideline for designing and exporting products to the United States and other Western countries and is available at:

https://www.cpsc.gov/s3fs-public/pdfs/blk_pdf_cpsia.pdf

Countries from the European Union also have laws on consumer product safety, rules on standards and risks, market surveillance rules, search for dangerous products. European Commission from the E.U. provides product safety and requirements information at:

https://ec.europa.eu/info/business-economy-euro/product-safety-and-req uirements_en

The International Organization for Standardization (ISO) provides the standard ISO 10377:2013, which was reviewed and confirmed in 2018. This standard deals with sources of practical guidance for suppliers to assess and manage the safety of consumer products, assessment documents, and risk management to meet applicable requirements. ISO assistive product standards are under discussion by the Technical Group 173. The Technical Committees of standardization agencies, such as ISO and CEN (European Committee for Standardization), are formed by professionals from a particular area. These professionals identify a need for **standard** in the market and begin to draft a solution for this demand. This proposal is then analyzed and voted on by professionals around the world. If there is consensus regarding its scope, definitions, and content, it becomes a standard. Other standards related to assistive technology and accessibility guidelines for disabled persons are available at the ISO Catalogue (ISO, 2019a).

This standard describes how to:

- Identify, assess, reduce, or eliminate hazards;
- Manage risks by reducing them to tolerable levels; and
- Provide consumers with hazard warnings or instructions essential to the safe use or disposal of consumer products.

The United States has provided disabled citizens with the Americans with Disabilities Act (ADA), which became law in 1990. The ADA gives civil rights protections to individuals with disabilities similar to those provided to individuals on the basis of race, color, sex, national origin, age, and religion. It guarantees equal opportunity for individuals with disabilities in public accommodations, employment, transportation, state and local government services, and telecommunications. Sangelkar and Mcadams (2012) explore the transferability of the knowledge contained in the Americans with Disabilities Act to universal product design. The authors used the International Classification of Functioning, Disability and Health (ICF) to formally describe user activity, the functional bases of design to describe product function, and the function-action diagram as a framework to create a detailed understanding of the interaction between a user and a product.

In the United Kingdom, the Consumer Protection Act 1987 is up to date with all changes known to be in force on or before 4 July 2019 (Legislation.gov.uk, 2019). The Act imposes new and more onerous demands on producers of goods with regard to safety. The general safety requirement means that it is a criminal offense to supply unsafe consumer goods in the United Kingdom. The Act states that: "a person should be guilty of an offence if he supplies consumer goods which are not reasonably safe having regard to all circumstances". These circumstances include: "the manner in which the goods are marketed and any instruction or warnings given with the goods, any published safety standards for those goods, the means, if any, and the cost of making the goods safer". So, meeting safety standards is essential for manufacturers to protect themselves against product liability.

Since many products are designed for children, some standards and guidelines are recommended relating to product safety for this group (Children's Products, 2020). Children's Products are consumer products designed or intended primarily for children 12 years of age or younger.

Attending to safety standards and regulations is an essential part of the design of a safe product. However, they just define the minimum requirements for safety. These kinds of standards and regulations may cover specific product attributes, testing procedures, or product performance. Ergonomic recommendations for designing products for children and adolescents are provided by Lueder and Rice (2008).

It is possible to reduce accidents by improving design through the implementation of safety standards. It is certain that as higher standards are enforced, there will be more pressure on manufacturers and designers to improve the quality and, above all, the safety of consumer products.

TIP

The second edition of the "Handbook on Standards and Guidelines in Human Factors and Ergonomics", edited by Waldemar Karwowski, Anna Szopa, and Marcelo M. Soares (2021), has already been released by CRC Press. The handbook contains chapters referring to standards on Human Factors and Ergonomics, nature of standards and recommendations, standards for evaluating work postures, standards for manual material handling tasks, standards for human–computer interaction, safety protection standards, ergonomics standards for the military area, and references on ergonomics standards.

3.8.3 Responsibility in Product Design and Manufacturing

Product liability usually results from the application of negligence, breach of warranty, or strict liability in tort[*]. Product liability law provides a formal mechanism for addressing issues related to product safety and resolving legal disputes involving injuries or death. It refers to legal action taken under tort in which an injured party (the plaintiff) seeks to recover damages for personal injury or loss of property from a commercial provider of a product in whole or part (seller, designer, manufacturer, distributor, etc.) because the plaintiff alleges that the injuries or loss resulted from a defective product (Kreifeldt, 2007; Sanders and McCormick, 1993).

Ryan (1985) discusses law cases in which courts have found manufacturers liable for injuries associated with the use of their products. The products in question conformed to existing safety standards but were found to be defective because they did not provide that degree of safety expected by the consumer. Hence, reaching minimum requirements for safety may not be enough. This is of vital importance, especially when referring to products destined for the international market that will have to meet the requirements of the laws of different countries.

---

[*] Tort. The term in common law systems for the civilly actionable harm or wrong, and for the branch of law dealing with liability for such wrongs (Abbott and Tyler, 2017).

TIP

Do you want to know more about the implications of introducing the product liability law in court for injuries caused by defective products? See: Owen and Davis (2019), Hunter et al. (2018), Abbott and Tyler (2017), Abbott (1980), Ottley et al. (2013), Soares and Bucich (2000), Dewis et al. (1980), Hunter (1992), and Wilson and Kirk (1980).

The growth in product liability cases has created a demand for ergonomics experts, both in the initial design to make products safer and in the courtroom. Ergonomists can play an important role in the court as expert witnesses, providing testimony to clarify technical issues. Karwowski and Noy (2005) address the application of ergonomics, knowledge, and techniques to standards of care in the context of regulatory and judicial systems. This activity is known as **Forensic Ergonomics**. It can help prevent the recurrence of system failures through administrative or engineering controls. The authors argue that the use of ergonomics in a preventive way can help in the perception of exposure to the responsibility of designer manufacturers regarding the safety of products, workplaces, and services provided.

ATTENTION

Forensic Ergonomics is an area of ergonomics specialized in providing technical and scientific testimony in courts in actions related to industry and consumer products. It can provide support in identifying responsibilities in labor claims, product safety, and service provision. Even with such a broad and potentially important impact, this specialty is still quite incipient in the world.

Want to know more about Forensic Ergonomics? See:
Handbook of Human Factors in Litigation (Noy and Karwowski, 2004).

Soares and Bucich (2000) described a case presented in court in the city of Rio de Janeiro, Brazil, in which the steps and procedures performed by a design and ergonomics specialist are described to provide a technical opinion on a product allegedly causing an accident. The product was a device for heating wax for hair removal. It was proven that during use, there were design flaws such as improper sizing, defects in raw material, incorrect manufacturing and assembly operations, insufficient thermal insulation, and electrical insulation failures, which resulted in accidents involving injuries to the user.

In addition to product safety, a range of product features constitutes the properties which will enable the product to meet users' needs. User needs will be discussed in the next chapter.

3.9   SUMMARY OF THE CHAPTER

- The product design process is a method composed of a set of rational and systematic procedures with the objective of conceiving and developing physical products to be employed by users.
- It is also a series of compromises between several product requirements: function, performance, reliability, usability, appearance, and cost. Finding the exact solution is sometimes very difficult, and a compromise solution has to be established between several acceptable solutions.
- The design process for consumer products can be summarized as consisting of six main sequential, or sometimes concurrent, phases: (a) Specification; (b) Conceptualization; (c) Modeling and Prototyping; (d) Product Evaluation; (e) Production; and (f) Marketing and Evaluation.
- The measurement of the interaction between consumers and products can provide requirements to improve the product's ergonomic specifications and general qualities. Only by such an approach can inadequate designs be identified.
- Usability tests are an important and essential part of the product development process. They are concerned both with obtaining user requirements prior to or initiating the product design process and in the early stages of design; and evaluating products when they have been built.
- The necessity to evaluate consumer products physically and ergonomically comes from the necessity of manufacturers to evaluate their new products and compare them with those already in the market, especially those of competitors.
- An extremely difficult problem faced by the manufacturers of products for disabled populations is that on the one hand, the manufacturers need to produce the highest possible volume of products to reduce manufacturing costs, while on the other, these products should suit the users' individual capabilities and limitations. Taking into account individual needs, the use of the universal design approach and modular design may be the solution to this problem.
- Attending to safety standards and regulations is an essential part of the design process.

4 Defining User Needs
A Tool for the Design of Competitive Products

This chapter presents the definitions of customer, user, and consumer, and aspects related to purchase preferences for specific products. Aspects related to product design for users with disabilities are also discussed. The concepts of needs, wants, product requirements, and specifications are presented. The importance to define product requirements as a specific guidance on how to design and engineer a product using measurable data is pointed out. The definition of "voice of the consumer" as a hierarchical set of "consumer needs" is presented and discussed. The chapter also addresses product requirements as a step to identifying user needs, including the concepts of user satisfaction and dissatisfaction.

4.1 DEFINING CUSTOMER, USER, AND CONSUMER

Usually, the words customer, client, and user are used synonymously. In this book, the **client** is defined as the person who buys the goods or services, and the **customer** or **user** is someone who effectively uses the product. It is important to observe that sometimes the client plays the same role as the consumer/user when he or she buys the product for his or her own use.

A user's preference for a specific product can be regarded as the result of a match between characteristics of the products, including design and style, and the product demands of the user. Of course, users (subject to certain constraints) will attempt to buy the product with the best match. This process of matching is a form of user information processing. It is influenced by several factors based mainly on the user's personal ability to differentiate, to discriminate, and to integrate information.

In a competitive marketplace, a user can generally choose freely to seek satisfaction among several distinct versions of the same product and is under no obligation to continue using this product.

ATTENTION

People have their own aspirations, values, and status symbols. Marketing strategies extensively explore these characteristics of human beings in order to sell products. However, marketing strategies have been applied very timidly to products for independent living. Although we know that many people with

DOI: 10.1201/9781003214793-4

disabilities buy their own products, many others are provided by entities, hospitals, clinics, or other government agencies, which, in most cases, prioritize cost.

With the global market breaking down boundaries, competition is increasing, and the figure of hundreds of thousands of disabled people is a segment in the marketplace to be considered by any company aiming for success with mass production.

It is common sense that the products designed for the disabled population have to meet the specific medical or therapeutic needs of that population. In fact, apart from the needs resulting from their disabilities, which are crucial, the needs of the disabled people are the same as those of the able-bodied population in terms of aspirations, uniqueness, values, and status. Problems arise when designers perceive disabled people as isolated sets of symptoms rather than whole people with needs for products that represent their lifestyle. The result of this can be inappropriate products which, in many cases, have their styling associated with medical and assistive products. These latter products frequently contribute to stigmatizing and can be rejected and abandoned even though they may be of clinical benefit. Products with these characteristics contribute to people with disabilities been perceived as in *need* and *surviving* and serve only to increase, at a psychological level, a person's sense of being disabled.

In this sense, some conditions are indispensable and must always be kept in mind when designing a product or service for people with or without disabilities, such as ensuring the right to come and go, to appreciate autonomy and independence, to provide conditions of use with comfort and safety, considering a wide range of users.

According to Barber (1996), if the design of products for independent living is purely to solve a problem based on the physical and medical needs of the people who will use them, the only values that will be reflected in the product are those of need and dependency. This, continues the author, betrays an underlying assumption that the people who rely on the products have no expectations in life beyond those of safety, security, and survival. Although these expectations are essential for all people, they should not be considered to the exclusion of the intended lifestyle, image, status, and identification. Such factors must be considered by designers and manufacturers, essential ingredients for the success of any product, however practical its purpose.

4.2   USER NEEDS AND REQUIREMENTS

Mitchell (1981) has defined **requirements** as those characteristics of a process, product, or place which should be provided to allow an individual to function effectively, safely, comfortably, and easily. If **requirements** are considered as the inputs from the environment which are needed by the user to permit optimum function, **demands**, on the other hand, represent the output that is needed by a user from particular equipment or environments. According to the author, requirements represent specific expressions of general needs.

Ulrich and Eppinger (2019) have distinguished **needs** and **requirements** as synonymous with **specifications**. They argue that **needs** are not specific to a particular concept and are independent of any product that may be developed, so designers should be able to identify user needs without knowing if or how they will eventually address those needs. For example, identifying the needs of the users of a wheelchair may include needs such as: a wheelchair that is lighter, easier to maintain, and cheaper. At this stage, the designer does not yet know whether he can provide a product that meets these characteristics. On the other hand, **requirements** do depend on the selected concept. These selected concepts are defined by, for instance, what is technically and economically feasible, by what competitors offer in the marketplace, and by user needs, as well.

ATTENTION

The concept of **needs** and **wants** has been defined, without consensus, by a number of authors, including Solomon (2016), Engel et al. (2005), and Mowen and Minor (1997).

4.2.1 General Considerations about User Needs

All user needs have to be met, and the product features should be responsive to those needs (De Feo, 2017). Observing user needs is today a powerful tool for the design of competitive products. User needs are requisites, desirable, or intrinsic, to be fulfilled by the product or service. Some have a higher priority for consumers than others. It is essential to identify consumer needs and establish priorities so that they can be useful to the engineering team during product development. Pugh (1991) says that any mismatches that arise between company products and the real needs of the consumer seem only be solved after the product is available on the market and after a long time of use.

TIP

A guide to identifying users' needs is found in the manual presented by Jan Dittrich.

A beginner's guide to finding user needs, Jan Dittrich, 2020
https://jdittrich.github.io/userNeedResearchBook/

Harris (1990) states that if, on the one hand, several companies are spending more money on trying to sell mediocre products, the successful companies are investing heavily in ergonomics and design and producing products that are more desirable to own. Examples of the latter approach can be seen in some smartphones currently on the market that excel in beauty and usability.

In this way, the importance of identifying a consumer need or wish to purchase a product and using market pull to sell the product should be considered. A long-term approval and acceptance of a product lie with users developing an affinity for the product and a belief that it meets their needs. A necessary condition for product success is that a product offers perceived benefit to the user, and it will do just that when it satisfies their needs (Ulrich and Eppinger, 2019).

According to Holt (1989), many designers concentrate their attention on technology and neglect the user's problems and needs. Identify the correct consumer needs, provide the designer with the potential to obtain feedback on the performance and acceptability of the design among users, and they enable the designer to make modifications that will improve the original design in terms of their requirements. If feedback is lacking, the design process runs the risk of becoming increasingly divorced from the reality of use. According to Harker and Eason (1984), designers need an effective way to represent user needs in the design process. Perhaps personas can help the designers see the person with disabilities more holistically. The product development process translates consumer needs on functional requirements into specific engineering and quality characteristics (Gryna, 2016).

In terms of the buying process, consumer needs presuppose expected benefits, which include an evaluation of alternative products before the purchase. An initial process of need recognition gives information that will permit the start of the evaluation of alternatives. For example, when buying a wheelchair, the user must consider the benefits that the wheelchair will provide, initially based on the information provided by the manufacturer and the salesperson. Only after the actual use of the wheelchair will the user know whether or not his expectations existing in the initial evaluation were met. This happens with the purchase of any consumer product.

According to Engel et al. (2005), there are three determinants of need recognition: (1) information stored in memory, (2) individual differences, and (3) environmental influences. The success of the alternative evaluation is defined only after the purchase, and it may result in the decision of buying or not buying a similar product next time. Furthermore, during this alternative evaluation, state the authors, it is not just the extent to which products meet expectations in terms of efficiency and effectiveness that determine user satisfaction. Still, other benefits of buying the product play an important role, for example, any discounts or accessory packages that are offered together with the product.

REFLECT

It is obvious that the user's needs vary from one to the other. Why do you think that in contemporary society, particularly in more developed countries, user needs are established more by social or emotional factors than by biological needs (food, shelter, etc.)?

Competitive markets have many products that are so much alike in practical or functional terms that a user's choice is often determined solely by the psychological perception of how the product will perform for them. Thus, products have both **denotative** (rational, functional, conscious level) and **connotative** (emotional, affective, unconscious level) aspects (Gregory, 1982). The author states that products are collections of meanings, and in order to be successful, it is necessary that they communicate satisfaction at both the rational and irrational levels. Consumers often choose products they associate with a certain lifestyle, believing that the qualities represented by the product image somehow correspond to their own or will somehow rub off onto them (Solomon, 2016).

Griffin and Hauser (1993) define the "voice of the customer" as a hierarchical set of "customer needs" where each need (or set of needs) has assigned to the product a priority, which indicates its importance to the customer. If listening to the "voice of the customer" during the design process seems obvious, it does not always correspond to reality.

INTERNET

In this TED talk, designer Timothy Prestero launches a manifesto for design to be used in the real world. To do this, the designers need to hear "the voice of the user" and understand the process of production and distribution of the product.

"Design for people, not awards", Timothy Prestero, June 2012.
https://www.ted.com/talks/timothy_prestero_design_for_people_not_awards
    ?language=pt#t-634687

4.2.2 PRODUCT REQUIREMENTS

Establishing **product requirements** is the next step after consumer needs have been identified. This procedure involves providing specific guidance on how to design and engineer a product using measurable data. Consumers usually express their needs using their own language. The design team needs to have this information in quantitative data as free of subjective interpretations as possible. In this way, **product requirements** can be understood as a set of specifications that translate consumer needs into precise and measurable data in order to produce products that are technically and economically realizable.

ATTENTION

Users often express their needs using their own vocabulary. So it is required to translate these needs into measurable data that can be used in the product manufacturing process. For example, if the user expressed that he or she would like to have a chair with a soft and comfortable seat, the measurable data would be to define the correct density of the foam that will make up the seat of the chair.

4.3 USER SATISFACTION AND DISSATISFACTION

Users express their satisfaction or dissatisfaction when using the product as a result of meeting or not meeting their needs (although the marketing literature distinguishes between dissatisfaction and non-satisfaction, in this book, we will use only the term dissatisfaction to denote users not satisfied with the product). **User satisfaction** can be defined as the user's response to the evaluation of the perceived discrepancy between prior expectations (or some other norm of performance) and the actual performance of the product as perceived after its consumption (Tse and Wilton, 1988). There is a growing managerial interest in user satisfaction as a means of evaluating quality and as a criterion for diagnosing product or service performance.

Satisfaction may be considered as a major outcome of marketing activity once it provides useful data in terms of post-purchase phenomena such as attitude change, repeat purchase, positive word-of-mouth, and allegiance to a brand (Churchill and Surprenant, 1982). The authors state that since the early 1970s, the volume of theoretical consumer satisfaction research has been increased impressively. Most of these studies have used some variant of the disconfirmation paradigm. According to the authors, an individual's expectations are (a) confirmed when a product performs as expected, (b) negatively disconfirmed when the product performs more poorly than expected (known as dissatisfaction), and (c) positively disconfirmed when the products perform better than expected.

In a model of satisfaction as a function of expectation and disconfirmation, Oliver (1993) points out that consumers are posited to form pre-consumption expectancies, observe product (attribute) performance, compare performance with expectations, form disconfirmation perceptions, combine these perceptions with expectation levels, and form satisfaction judgment. Therefore, this model postulates that satisfaction acts as a mediator between pre-exposure and post-exposure attitudes.

We can think of this model of user satisfaction as a continuous line (Figure 4.1): on one side, we have **Total Satisfaction** and, on the other side, **Total Dissatisfaction**. Between one and the other end of this line, we have partial levels of satisfaction or dissatisfaction. That is, the user may be totally or partially satisfied or dissatisfied with a product.

Mano and Oliver (1993) characterize **product satisfaction** as "an attitude-like post consumption evaluative judgment with the evaluative aspect of that judgment varying along with the hedonic (pleasantness) continuum" (p. 451). De Feo (2017) argues that product satisfaction and product dissatisfaction are not opposites. The author states that (a) the first one has its origin in product features and is why clients buy the product; (b) the second one has its origin in nonconformances and is why users complain.

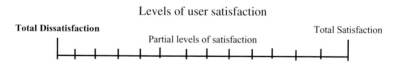

FIGURE 4.1   Model of user satisfaction. Source: The author.

Users are usually engaged in evaluating products that they use in their daily activities. **User satisfaction/dissatisfaction** is determined by the overall feelings or attitudes he or she has about a product after it has been purchased (Solomon, 2016). In this way, product satisfaction and product dissatisfaction are two important concepts to be understood.

Thus, we can summarize satisfaction and dissatisfaction as follows (De Feo, 2017):

- **Product satisfaction** occurs when product features fully reflect user needs.
- **Product dissatisfaction** occurs when products have deficiencies, and its characteristics do not reflect user needs fully.

ATTENTION

In terms of product design, **consumer satisfaction** is compounded with visual appeal, "feel", functionality, expectations, and aesthetics. Indeed, a successful design needs to consider these aspects altogether. Concentrating on any one aspect to the detriment of another may cause dissatisfaction. In terms of products designed for disabled people, the medical and therapeutic characteristics of the product are part of its functional features, which, in conjunction with the other product's features, must meet user needs and contribute to express his or her satisfaction.

All of this reflection is important for product design, as well as for broadening the designer's vision of users' diversity and complexity. This applies to a better understanding of the needs of some specific populations, such as the elderly, people with disabilities, pregnant women, and obese people, whose needs still need to be the subject of greater attention, research, and be properly met by designers. It is important to keep in mind that considering products designed to meet the specific needs of users is to create opportunities for their involvement in daily activities and foster social participation.

The next chapter analyzes the currently available methods in the design and manufacture of products based on user needs.

4.4   SUMMARY OF THE CHAPTER

- Apart from needs resulting from their disabilities, which are crucial, the needs of the disabled population are the same as those of the able-bodied population in terms of aspirations, uniqueness, values, and status.
- Marketing strategy explores widely the consumer product characteristics which aim to meet people's aspirations, values, and status symbols. However, marketing strategies have been applied very timidly to products for independent living. This is mainly because the main consumer is government agencies and not the actual people who use the product.

- Investigating consumer needs provides the designer with the potential to obtain feedback among users on the performance and acceptability of the design. The feedback enables the designer to make modifications that will improve the original design.
- As a result of meeting or not meeting their needs, users will express their satisfaction or dissatisfaction with the product.
- In terms of product design, consumer satisfaction is compounded with visual appeal, "feel", functionality, expectations, and aesthetics. Concentrating on any one aspect to the detriment of another may cause dissatisfaction.
- In terms of products designed for the disabled population, the medical and therapeutic characteristics of the product are part of its functional features, which, in conjunction with the other product features, must meet user needs and contribute to consumer satisfaction.
- The design of products for disabled people must take into consideration, besides medical and therapeutic aspects, aspects such as aspirations, uniqueness, values, and status specific to this population.

5 Design for Disabled People

Analyzing the Needs of a Special Population

This chapter begins by presenting data and considerations about the population with disabilities. The definition of disability and some associated concepts are introduced. An analysis of the design and product requirements that include people with disabilities is also presented. A discussion is conducted on the absence of data on the capabilities and limitations of the disabled that assist in product design. The design for the disabled and the universal design is presented and discussed in detail. Finally, the classification and characteristics of products for people with disabilities are discussed and considered in product design for users with and without disabilities.

5.1 GENERAL CONSIDERATIONS ABOUT THE DISABLED POPULATION

An increasing segment of the world population is being reported to have some disability. According to 63 surveys conducted in 55 countries by the *United Nations Disability Statistics Database* (2020), the prevalence of disability amongst the population reported in various countries varies between a low of 0.4% in Qatar to a high of 35.2% in Sweden.

REFLECT

It is difficult to determine the exact number of people with disabilities or with limitations attributable to aging. There is also a greater risk of disability in old age, and the population in six continents is getting older at unprecedented rates. In reality, the estimated numbers of the disabled population can sometimes be inaccurate, considering people with multiple disabilities and in various classifications. Estimates vary considerably depending on the definition of limitation used. However, the figures expressed above are high enough to ensure that a very significant part of the population needs specific treatments (for example, provision of equipment, caregivers, or both) with the power to develop their skills and abilities and thus obtain a good quality of life. This is a huge market to be considered by the manufacturers of this equipment and its designers.

The serious impact the very large population of disabled people has on mass-market products is beginning to be recognized by manufacturers. Also, many governments are becoming more aware of the problems arising out of this large number.

Apart from humanitarian concerns, the costs to society that result from the need for special assistance required by disabled persons (unemployed or non-independent living) are very high. In addition, people with disabilities find it difficult to occupy jobs suitable to their potential. Often, they are not taken advantage of in their real capacities, interfering in their productivity. Hence the need to think about products and services that will contribute both to good work performance and to the daily activities of this population.

It must also be considered that a significant portion of the disabled population lives at or near the poverty level, mainly in poor and developing countries. Kumar (2009) states that the low employment among persons with disabilities is not unique to a few countries but is a general feature across most countries. Even if the disabled person is employed, concludes the author, he or she makes far less money than their non-disabled counterpart. These facts represent a social problem of a very large amplitude.

NOTE

One billion people, or 15% of the world's population, experience some form of disability, and the prevalence of handicap is higher in developing countries. A fifth of the estimated global total, or between 110 million and 190 million people, experience significant forms of disability (The World Bank, 2020).

5.2 DEFINING DISABILITY AND ASSOCIATED CONCEPTS

The understanding of disability has been culturally and socially constructed. At each historical moment, it presents itself differently, depending on culture, access to information, and beliefs or convictions.

The concept of people with disabilities has evolved over the years, undergoing some changes, and several definitions for the same term can be found. According to the International Classification of the Functioning of Disability and Health of the World Health Organization (WHO, 2005), disability is any permanent or temporary loss or abnormality of a psychological, physiological, or anatomical structure or function. It includes the existence or appearance of an anomaly, defect, or loss of an extremity, organ or body structure, or a defect in a functional system or mechanism of the body.

The current nomenclature Person with Disability (PwD) was adopted from the United Nations Convention on the Rights of Persons with Disabilities in 2006. Since then, it has been agreed that when referring to these people, we should use this term.

Physical or mental limitations can be classified on three levels: (1) Impairment, (2) Disability, and (3) Handicap (based on Kroemer et al., 2018; Pirkl, 1994; Soede, 1990; and Nichols, 1976).

- **Impairment**, a result of diseases or accidents, is characterized by losses or abnormalities in part or all of a limb, organ, tissue, or other body structure, including the systems of mental function. Thus, impairment is a medical or clinical disability, e.g., hearing loss, stiffness of a joint, or loss of a limb.
- **Disability** is any restriction or lack of functional ability, resulting from an impairment, to perform an activity in the manner or within the range considered normal for a human being, e.g., communication problems due to hearing loss or mobility problems due to joint stiffness; the man or woman who has had a leg amputation and uses a prosthesis (e.g., an artificial leg) does not have the same ability to move with the same agility as with two limbs considered healthy.
- **Handicap** is a disadvantage for a given individual, resulting from an impairment or disability that limits or prevents the fulfillment of a role that is normal (depending on age, sex, and social and cultural factors) for that individual. A handicap can also be considered as the social and economic disadvantage that results from impairment or disability. Again taking the example of somebody who had had a lower limb amputated, the disadvantage of having an amputated limb will depend entirely upon the patient's age, his job, where he or she lives, and the family situation.

Thus, the individual may have an impairment that causes a hearing loss. As a consequence, the impairment can cause deafness, which characterizes an inability to communicate or disability. Handcap would then be the absence of communication.

After several reviews, in 2001, the World Health Assembly approved the International Classification of Functioning, Disability and Health (ICF). The ICF is therefore based on a biopsychosocial approach that incorporates health components at the bodily and social levels. In the assessment of a person with a disability, the ICF model stands out from the biomedical one, based on the etiological diagnosis of the dysfunction, evolving into a model that incorporates the three dimensions: biomedical, psychological (individual dimension), and social. In this model, each level acts on and suffers the action of the others, all of which are influenced by environmental factors (physical and social), by different cultural perceptions and attitudes towards disability, by the availability of services and legislation, which may consist of barriers or facilitators for the individual's functionality (WHO, 2020).

The ICF describes functionality and disability related to health conditions, identifying what a person "can or cannot do in their daily life", in view of changes in the functions of the organs or systems and structures of the body, as well as the activity limitations and restrictions on social participation in the environment where he lives.

There are some difficulties in using the terms impairment, disability, and handicap consistently. The *Census Bureau of the U.S. Department of Commerce* controversially considers, in its *Statistical Abstract of the United States* (as cited in Elkind,

TABLE 5.1

Classification of disabilities according to Elkind (1990)

Types of disabilities

Sensory	Visually handicap
	Hearing handicap
Motor	Orthopedic impairments
Cognitive	Specific learning disabilities
	Speech impairments
	Mentally retardation
Illiterate and semiliterate	
Emotionally disturbed	
Other health impairment	

1990), limited literacy as a kind of disability. Arguments against this classification consider that this condition is not a physiological or psychological disability but something that can and should be remedied throughout education. The author introduces a classification based on the categories used by the *Census Bureau* to collect data about disabilities. This classification is given in Table 5.1.

It is worth calling attention to the population of elderly users, some of whom have great disabilities and physical and cognitive disabilities for some tasks, in the development of functions and in the use of many products and services. Although this population is not the subject of this book, it should be noted that their needs can often be similar to the needs of the disabled.

Hale (1979) controversially points out that physical disability, no matter how or by what it was caused, is a medically determined fact that can be defined and described explicitly. Some disabilities may seriously handicap a person in one situation and not in others. A concert pianist, for instance, who loses a finger, is seriously handicapped concerning his or her career but may be able to perform most other usual activities. Another important factor to be observed, according to the author, is that sometimes a handicap can be minimized or even completely eliminated (a) through the use of suitable aids, e.g., equipment, devices, and aids in general; (b) a friendly environment at home, work, transport, and in public places; and (c) through constructive and realistic attitudes. Sometimes, states the author, a person's view of his or her disability is more handicapping than the disability itself. It is curious that Hale does not mention help from other people.

A contrary point of view to that of Hale is put forward by Vanderheiden (1990). He states that it is important to say that there is not a clear line of demarcation between people who are categorized as disabled and those who are not. Considering if a certain performance or ability is under focus, what can be observed is a distribution including a small number of individuals who have exceptionally high ability, a large number of individuals with mid-range ability, and another long tail representing individuals with little or no ability in that particular area.

Few individuals are at the extremes of the distribution. That is, few are entirely incapable or completely capable. Thus, most of the people are distributed in positions between the extremes of the distribution, where their position will depend on their skills. In this way, a person who performs poorly along a distribution in one dimension (e.g., vision) may perform excellently with regard to another dimension (e.g., hearing or I.Q.). Therefore, continues Vanderheiden (1990), very few individuals are found at either end, at the extremes of the distribution, most of the population is located in different positions, depending on their specific capacity, distributed along this line.

On the one hand, we can exemplify the swimmer Michael Phelps, who won 16 Olympic Gold medals when he was at the peak of his physical condition. At the other end of the curve, we can exemplify with the English physicist Stephen Hawking, with a degenerative neuromotor disease known as amyotrophic lateral sclerosis, who, despite his physical condition, was one of the most outstanding scientists in history.

So, the distinction between "able" and "disabled" people may not be simple since it involves a continuous function rather than a simple dichotomous "able-disabled" distinction. People can be considered "temporarily non-disabled" because human beings experience temporary and/or functional limitations during their lives (through, for example, illness and accident and through the natural processes of development and aging).

Figure 5.1 presents a graph showing a distribution of the population in three categories: "fully non-disabled people", that is, athletes who are at the peak of their physical conditions and represent a minimum portion of the population; "people without disabilities", considered as the majority of people within a population, in an extract considered normal; and the "completely incapacitated people", who are at the

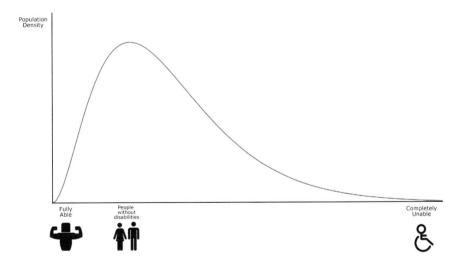

FIGURE 5.1 Distribution of the population among completely non-disabled and disabled people in the general population. Source: The author.

other end of the population, as in the example we mentioned above from the English physicist Stephen Hawking. It is very rare to identify people who are at the ends of either side of the curve.

5.3   THE DESIGN OF PRODUCTS WHICH INCLUDE THE DISABLED

Consumers, in general, are comprised of an extremely heterogeneous population in terms of physical and mental ability. Differences in age, size, shape, weight, etc., for both non-disabled and disabled persons, make designing products to satisfy the whole range of such diversity practically impossible. With handicapped people, their handicap exacerbates individual differences, which makes it more difficult for the designer.

The use of adequate aids, sophisticated or simple, can transform the daily life of the disabled. As a function of the actual stage of technology, many products that, some time ago, seemed possible only in science-fiction are now a reality.

On the other hand, many devices that are readily available are standard, inexpensive, and, in some cases, tailor-made for one individual. From a simple home-made reacher that retrieves dropped articles to a sophisticated breath-controlled switch that operates a computer keyboard, people with disabilities can find a large range of products that aid them in living a more independent life.

INTERNET

What technologies are available to help people with disabilities? Numerous high-tech products to help the disabled are available on the market, from software to home robots.

Read more at:

Technology is most powerful when it empowers everyone, Apple, 2020
https://www.apple.com/accessibility/
The power of technology for people with disabilities, Microsoft, 2020
https://news.microsoft.com/features/the-power-of-technology-for-people-with
-disabilities/
What are some types of assistive devices, and how are they used? U.S. Department of Health and Human Service, 2020
https://www.nichd.nih.gov/health/topics/rehabtech/conditioninfo/device

Thanks to the advance of science and technology, people can expect to live longer, recover more fully from illness, and lead active lives in spite of trauma. When it comes to people with disabilities, if on the one hand, we have miraculous surgical techniques, bioengineering, and revolutionary medicines, on the other hand, we still have crutches, wheelchairs, and stair lifts that seem to come from a different age or civilization than high-tech equipment.

Generally, products are merely extensions of a human being's abilities. For example, the telephone simply lets people talk over longer distances; the airplane allows people to cover greater distances more quickly; a computer allows people to calculate quickly and accurately; a pencil allows people to communicate thoughts without having to speak them and gloves will allow people to use their hands in extreme temperatures. Following this thinking, spectacles let people read and see more clearly; a hearing aid makes the sound more audible; a prosthesis (e.g., an artificial arm) allows a person to use a hand, and a wheelchair allows the user to "walk". The virtue of this line of thought is to eliminate the status of "special product" and de-stigmatize what are now referred to as prosthetic devices: just as the latest thinking of a spectacle user is that he or she is wearing an orthotic device. By way of clarification: a prosthesis replaces a lost body part, and an orthosis supports function and assists in the performance of a task.

The use of everyday products can turn an individual disability into a handicap. There are two approaches to tackle these problems:

- By adaptation of existing products and the development of special aids or
- By taking into account the limitations and capabilities of the disabled in the design of new products.

Disabled people represent a significant proportion of consumers in terms of their buying power. But, in fact, when a product is specific to a special segment of the disabled population, the economic buying potential decreases substantially and consequently may not receive enough consumer market attention. On the other hand, disabled people have difficulties in using effectively or safely standard consumer products because of their impairments.

The design for disabled people aims to design assistive products. According to the World Health Organization, assistive devices and technologies are those whose main objective is to maintain or improve the functioning and independence of an individual to facilitate participation and improve general well-being (WHO, 2020).

INTERNET

Watch the excellent video on design for people with disabilities:

When we design for disability, we all benefit. Elise Roy TED Class, 2015.
https://www.ted.com/talks/elise_roy_when_we_design_for_disability_we_all
    _benefit?referrer=playlist-designing_for_disability#t-131

5.3.1 DATABASES ON DISABLED CAPACITIES AND LIMITATIONS

If, on the one hand, a large amount of data about the capacities and limitations of different groups within the non-disabled and non-elderly members of the general population are available, on the other hand, data related to people with disabilities are still rare and urgently required. However, if designers wait until statistically representative data are available, many products that turn a person's disability into a handicap will certainly continue to proliferate. Indeed, rarely has the industrial designer had sufficient information available to him or her related to the various handicaps and how they would be affected by the design that he or she is working on. Disabilities and functional limitations of aging are now more frequently cited in textbooks and included in data tables than they were in the past. However, few designers research specific dissertations, theses, and data to find out when they will develop their products.

TIP

A database containing information useful for product design for people with disabilities and physical limitations should include data on Kumar (2009):

* strength, endurance, and range of motion;
* people's capabilities in standard activities such as pinching, gripping, lifting, pulling, and pushing from the point of view of strength and also the ability to sustain them;
* motion at different body joints (upper extremities, trunk, head, neck, and lower extremities); and
* balance, stability, capabilities of sight, and hearing.

5.3.2 DESIGNING FOR PEOPLE WITH AND WITHOUT DISABILITIES

Although making compatible products for both disabled and non-disabled people has been a very difficult task, products usable by disabled consumers will usually be well-accepted among a portion of the non-disabled and aging population, especially if those products do not carry with them a stigma of handicap. Products designed for those with disabilities, keeping the non-disabled in mind, and vice versa, would nicely avoid the standard marketing problem of segmenting the handicapped from the non-disabled, asking how many handicapped there are, and the reaction that the market is too small to address.

REFLECT

Do you know of a product that was designed to include the needs of users with and without disabilities? Do you think the product serves the two target audiences well?

Producing products that can be used by the largest number of users possible, including the disabled and the aging, is both an economic and social strategy that helps to contribute to product success. This is true because enlarging markets products become less expensive than lower-production "specials products".

Vanderheiden (1990) advocates that in some cases creating a design that is more accessible to both the non-disabled and disabled contribute to (a) decreasing the costs involved in manufacture or maintenance/support of a product (e.g., signal tones and light in a lift, in advance of its arrival at the floor, has solved accessibility problems without increasing costs) and (b) increasing the functionality for non-disabled users including benefits such as lower fatigue, increased speed of operation, and lower error rates (e.g., lighter hand brake, television subtitles, etc.).

The investment in research and development for some disability products is sometimes disproportional to the return from the product. To make the product economically viable other markets may need to be found. In this way, the challenge for designers is to elaborate design specifications to integrate the needs and characteristics of special populations as part of a broader population.

There are several different terms to define the design concept for as many people as possible (see Glossary in Note below). In fact, all these terms are sound common sense within the context of the "ergonomics approach".

It is important to be clear that although the elderly and the disabled people should be included in the design process, it is not possible to design all products and devices so that they are usable by all individuals. There will always be a segment of individuals who are unable to use a given product.

Because of the diversity of disabilities, the number of individuals with any one particular type or combination of disabilities is much smaller compared with the population as a whole (Vanderheiden, 1990). In spite of this, states the author, it is more difficult to accommodate this population in the overall design process because of the many dimensions that need to be considered. Furthermore, in the same way that, economically speaking, it is unreasonable to design everything to be usable by everyone, it is equally unreasonable to produce special designs for each major consumer product to accommodate the different disability groups. Some special aids and other devices will continue to be necessary to fulfill those needs that accessible mass-market design cannot effectively meet. Based on this principle, the term "universal" would be conceptually mistaken when we speak of Universal Design.

ATTENTION

SOME DEFINITIONS

Design for All is defined as that which aims to allow all people equal opportunities to participate in all aspects of society, i.e., everything that is designed for one person should be used by the entire population and be well accessible, convenient to be used by all in society and appropriate to the evolution of human diversity (Persson et al., 2015).

Universal design is the design of products and environments that can be used by all people, as far as possible, without the need for adaptation or specialized design (Mace et al., 1991).

Inclusive design is the design of products and/or services conventionally accessible and usable by as many people as reasonably possible (...) without the need for special adaptation or specialized design (British Standards Institute, 2005).

Accessible design is the design focused on the principles of extending the standard design to people with some kind of performance limitation to maximize the number of potential customers who can readily use a product, building, or service (ISO, 2001).

So, the best and most economical approach appears to be making mass-market goods more accessible through a design that carefully tries to include disabled and non-disabled altogether, respecting their differences and necessities. The ergonomic methodology proposed here is a suitable tool to develop such a design.

The Center for Excellence in Universal Design (NDA – National Disability Authority, 2020) points out the Seven Principles of Universal Design:

1. **Equitable Use** – The design is useful and marketable to people with diverse abilities.
2. **Flexibility in Use** – The design accommodates a wide range of individual preferences and abilities.
3. **Simple and Intuitive Use** – The use of the design is easy to understand, regardless of the user's experience, knowledge, language skills, or current concentration level.
4. **Perceptible Information** – The design communicates necessary information effectively to the user, regardless of ambient conditions or the user's sensory abilities.
5. **Tolerance for Error** – The design minimizes hazards and the adverse consequences of accidental or unintended actions.
6. **Low Physical Effort** – The design can be used efficiently and comfortably and with a minimum of fatigue.
7. **Size and Space for Approach and Use** – Appropriate size and space is provided for approach, reach, manipulation, and use regardless of user's body size, posture, or mobility.

Vanderheiden and Vanderheiden (2019) point out four different approaches to make products more accessible in order of desirability. It is important to observe that it may be necessary to use one or a combination of these approaches to achieve the desired level of accessibility.

- **Direct accessibility**, corresponds to producing modifications, incorporated into the initial product design phase, which can significantly increase accessibility and usefulness to individuals with functional impairments; e.g., Mouse Keys accessible options on Apple computers that allow the user to move the cursor across the screen using the numeric keypad rather than the mouse.
- **Accessibility via standard options or accessories (from the manufacturer)**, means to provide adaptations or alternatives to standard design when it is not possible to design the standard product to make it directly accessible for some disability populations; e.g., availability of microwave ovens control panel with ridges around each button and some type of tactile identification of button function to replace the usual buttons which are not easily distinguishable by touch.
- **Compatibility with third-party assistive devices**, means the establishment of cooperation between mass-market manufacturers with assistive device manufacturers facilitating efforts of third-party manufacturers in a number of ways, including using standard approaches, providing appropriate connection points, advance access to new versions of products, and technical assistance in understanding and properly attaching accessories to the product; e.g., keyguards and accessories to keyboards and providing compatibility between standard computers with alternative input devices to fit people with a variety of severe physical disabilities.
- **Facilitation of custom modifications**, when all the other approaches prove to be impractical or uneconomical, the best solution may be to carry out custom modifications of the product, e.g., adaptations of automobiles used by drivers with physical impairments.

According to Feeney and Galer (1981), the main difficulties in finding generalizable ergonomics solutions to the problems presented by disability, in all its many varied forms, are related to goals, classification, and measurement.

5.3.2.1   *Goals*
The goals are based on two approaches:

a) the first one states that people who are physically impaired are different in their capacities or characteristics and consequently need special arrangements to use standard equipment that is designed for the non-disabled population. Hence, many gadgets and adaptations are on the market to enable those with impairments to use standard equipment which is designed for the non-disabled population; and

b) the second approach points out that, if when designing products and environments for all users, the requirements and capacities of impaired people are incorporated into design solutions, the need to use special aids and adaptations would disappear and impaired people would be better integrated into society. Certainly, some limits of this approach must be observed since the blind, the deaf, and those who use wheelchairs will always need special care. On the other hand, handicapped people with more common functional impairments may take advantage of a "universal design" approach. But it is obvious that certain members of the impaired population are so limited in their capacities that any design solutions must be tailored to their individual requirements.

5.3.2.2 *Classification*

Classification of impairment is described in medical terms and, although adequate to identify and prescribe medical and therapeutic treatment procedures, does not provide a basis for the assessment of physical and mental abilities, which can be used by the ergonomist and designer.

5.3.2.3 *Measurement*

Measurement and analysis of body dimensions of disabled people are extremely difficult because they present skeletal deformity and variation so that reference points, usually applied to the general population, are inappropriate, and the variations in stature and shape are impossible to manage.

In terms of design for the disabled, a strategy points out that people with different disabling conditions have difficulty performing similar tasks because of his or her disability. In this case, it can be suggested that by identifying common aspects of difficulties of task performance, generic solutions that may be applied to overcome parts of an individual's impairments will also help other individuals with different impairments.

The implications of identifying generic solutions to the designer and manufacturer are that a large market can be identified, more economical manufacturing processes used, and investment in research and development might be more possible. Kumar (2009, p. 30) concludes that "given the size and significance of the population with disability (due to aging, trauma, or disease), the rationale of extensive application of rehabilitation ergonomics is not only economically viable but profitable".

Cushman and Rosenberg (1991) point out that, in general, design solutions, including the disabled, fall into four categories:

- **Improving access to displays and controls**, e.g., increase the size of the lettering on displays and labels, use displays with high contrast and a wide viewing angle, place the control panel on the front surface of the product, etc.
- **Simplifying product operation**, e.g., make the operation of the product self-evident, minimize cognitive demands by providing appropriate task aids (such as adequate labels, operational sequence diagrams, and pictograms) and simplify user manuals, etc.

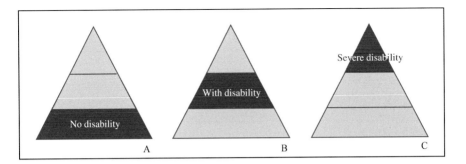

FIGURE 5.2   The "User pyramid" showing levels of disability. Source: The author.

- **Providing redundancy for sensory information**, e.g., use both visual and auditory displays to convey the same information, provide redundancy for coded information (such as simultaneous use of color coding and brightness coding), and provide several types of feedback – visual, auditory, and tactile – when feasible, etc.
- **Tailoring the product to meet the needs of the individual user**, e.g., provide the capability for adding prosthetic devices to meet the needs of specific individuals (such as image enhancers, speech synthesizers, headphones, touch screens), provide brightness, contrast, and loudness controls and provide the capability for users with deficiencies in color vision to select colors when any type of color coding is used.

As has been cited many times throughout this book, the inclusion of users into the very beginning of the design process is essential. Taking the Universal Design approach into account, understanding non-disabled and disabled users' abilities and limitations is a crucial starting point in the development of a new product.

Figure 5.2 shows the "User Pyramid" representing all users in their daily activities and the incidence of disability of varying severity that affects them. At the base of the pyramid (A) are non-handicapped and older users with slight disabilities, e.g., some deterioration in strength, sight, or hearing). In the middle part (B) are people with more severe disabilities due to illness or age, e.g., people who need aids – wheelchairs, some special equipment, etc. – to perform activities considered normal for a human being. At the top of the pyramid (C) are people with severe disabilities, e.g., people with very little strength or mobility in arms or hands.

5.4   CLASSIFICATION AND CHARACTERISTICS OF PRODUCTS FOR DISABLED PEOPLE

5.4.1   CLASSIFICATION OF PRODUCTS FOR THE DISABLED

ISO 9999 (2016) has established a classification and terminology of assistive products, especially produced or generally available, for persons with disabilities. Assistive products used by a person with a disability, but which require the assistance

TABLE 5.2

The classification system for aids for disabled persons (Research Institute for Consumer Affairs, 1984)

Types of aids	Examples
Aids for therapy and training	Aids for inhalation, circulation, and dialysis therapy, stimulators, aids for continence training, etc.
Orthoses and prostheses	Prostheses of upper and lower limbs, orthopedic shoes, etc.
Aids for personal care including clothes and shoes	Aids for toileting, thermometers, barometers, personal scales, etc.
Aids for transportation and locomotion	Walking aids, car adaptations, mopeds, cycles, wheelchairs, mobile patient lifts, orientation aids, etc.
Household aids	Cooking aids, dishwashing aids, aids for eating and drinking, housekeeping aids, sewing and mending aids, etc.
Aids for adaptation of homes and other premises	Tables, light fixtures, chairs, beds, support devices, door and window openers/closers, safety equipment, etc.
Aids for communication, information, and signaling	Braille and similar systems; manipulators and robotic arms; electric-optical aids; writing, reading, and drawing aids; telephonic aids; audio-video aids; hearing aids; alarm systems; etc.
Aids for the handling of other products	Package openers; extenders; forehead, chin, and mouth stick; remote control aids; push-bottoms; knobs; latches and handles; grips and holders, etc.

of another person for their operation are included in the classification. The Research Institute for Consumer Affairs (1984) has established a classification of aids for people with disabilities that has become a standard within the European Community. Table 5.2 shows the classification and some examples.

5.4.2　Characteristics of Products for the Disabled

Kroemer et al. (2018) point out some characteristics of assistive devices which can also be considered in product design for other users:

- **Affordability**, the extent to which the purchase, maintenance, and repair causes financial hardship to the consumer;
- **Dependability and durability**, the extent to which the device operates with repeatable and predictable levels of accuracy for extended periods of time;
- **Physical security**, the probability that the device will not cause physical harm to the user or other people;
- **Portability**, the extent to which the device can be readily transported to and operated in different locations;

- **Learnability and usability**, the extent to which the consumer can easily learn to use a newly received device and can use it easily, safely, and dependably for the intended purpose;
- **Physical comfort and personal acceptability**, the degree to which the device provides comfort, or at least avoids pain or discomfort to the user so that the person is attracted to use it in public or private;
- **Flexibility and compatibility**, the extent to which the device can be augmented by options and to which it will interface with other devices used currently or in the future;
- **Effectiveness**, the extent to which the device improves the user's capabilities, independence, and objective and subjective situation;
- **Ease of assembly and maintenance**, the attribute of a product not to demand excessive strength, over-exertion or difficulties of understanding in assembly and maintenance; and
- **Ease of repair**, the availability of suppliers, spare parts and accessories, and facility of customer repair or supplier repair.

TIP

Where to learn more:

Constantine Stephanidis. (2009). *The Universal Access Handbook* CRC Press Taylor & Francis Group. ISBN: 978-0-8058-6280-5

William Lidwell, Kritina Holden and Jill Butler (2010). *Universal Principles of Design*. Rockport Publisher, ISBN-13: 978-1592535873

Soares, M. M. (1999). *Translating user needs into product design for disabled people: a study of wheelchairs*. Loughborough University, UK. Ph.D. Thesis. Available at: https://repository.lboro.ac.uk/articles/thesis/Translat ing_user_needs_into_product_design_for_disabled_people_a_study_of _wheelchairs/9356108/1

J.J. Pirkl (1994). Transgenerational design. Van Nostrand Reinhold. ISBN-13: 978-0442010652

5.5 SUMMARY OF THE CHAPTER

- One billion people, or 15% of the world's population, have some form of disability. A fifth of the global population, or between 110 million and 190 million people, are considered to be people with significant disabilities.
- The serious impact the very large population of disabled people has on mass-market products is beginning to be recognized by manufacturers internationally.
- When a product is specific to a special segment of the disabled population, the economic buying potential decreases substantially and consequently may not receive enough design attention.

- Differences in age, size, shape, weight, etc., for both non-disabled and disabled persons, make designing products to satisfy the whole range of such diversity practically impossible.
- However, designing products for people with disabilities, keeping the non-disabled in mind, and vice versa, while respecting their differences and needs, would avoid the problem of standard marketing and segmentation in the product market for the disabled and non-disabled.
- Products usable by disabled people are often well-accepted by the non-disabled population because they are easier to use, especially if those products do not carry a stigma of handicap.
- Although elderly and disabled people should be included in the design process, it is not possible to design all products and devices so that they are usable by all individuals.
- The ergonomics approach to wheelchair design must include a variety of techniques from physiological to behavioral assessment.
- Both the approach of the ergonomist and that of the industrial designer will contribute to meeting user needs and should include investigations on (a) data related to body dimensions, physical workload, functional demands, posture, subjective evaluation, and product safety; (b) tasks including transferring, driving, sitting, braking, folding, and loading and (c) the environment in which the wheelchair will be used.

6 Models and Methods Based on User Needs for the Design and Quality of Products

This chapter presents models and methods based on the users' needs for the design and quality of consumer products. Initially a discussion on new forms of production and quality is conducted. Aspects such as Industry 4.0, Society 5.0, Quality and its categories, customer satisfaction and needs are presented. Special attention is given to two new methods in the product design and manufacturing process: the Quality Function Deployment (QFD) and *Kansei* Engineering. Both are discussed in detail. The definition of the human–product interface is presented, as well as the definition, analysis, models, and tests of usability. Considering that the user experience (UX) is the result of the user interaction with the product or system (interface) the final part of this chapter is dedicated to the presentation and analysis of UX and usability. Aspects such as emotion, feeling and methods and tools for usability and user experience evaluation, including emerging technologies based on neuroscience and infrared computerized thermography, are discussed in detail. Finally, considerations of the contributions of design and ergonomics during and after the pandemic period including recommendations for improving the physical and mental health of individuals are presented.

6.1 GENERAL CONSIDERATIONS

Times have changed and, consequently, products have also changed. In practice, products move on a continuous spectrum from traditional to modern. For example, the advent of microwave cooking has changed the food industry from traditional to contemporary. The earliest automobiles and telephones were traditional in simplicity but now are modern in complexity. The change from traditional to modern is often gradual and can mask the need for new approaches in product development, including additional manufacturing and managerial methods and technological tools.

Manufacturing is generally defined as the conversion of raw materials, generally in a large-scale operation, into products. The basis of modern manufacturing is to accomplish this conversion with ease, quickness, and economy. Quality is a powerful tool used by companies worldwide to guarantee the strength of their products so that they remain competitive.

DOI: 10.1201/9781003214793-6

Nowadays there is talk of Industry 4.0, which aims to integrate systems such as the Internet of Things, Artificial Intelligence, cloud storage, cyber-physical systems, and cognitive computing, all chained between machines, devices, and systems controlling each other. In Industry 4.0 we have the chaining of machine–machine systems. The human touch and the return to the human–machine system take place one step further: Society 5.0 seeks the reconciliation of the human with the machine.

REFLECT

Have you ever thought about the possibility of a machine-dominated world? Do you think that in a world with advanced systems such as virtual reality, Big Data, remote sensing, autonomous equipment, wearable devices, interfaces between brain and computer, nanotechnology, and exoskeletons, this would be possible? Do you think that the 5.0 society can be a way to prevent the eventual dominance of the machine over most human activities?

INTERNET

Want to know more about Industry 4.0 and Society 5.0? Watch the videos and read the text below.

The next manufacturing revolution is here, Olivier Scalabre. TED talk, 2016.
https://www.youtube.com/watch?v=AyWtIwwEgS0
Realizing Society 5.0. Japan Government, 2020.
https://www.japan.go.jp/abenomics/_userdata/abenomics/pdf/society_5.0.pdf

As was cited previously in this book, the concept of **quality** adopted here is a user-based one, as defined by Juran (2016). It consists of specifying those product features which meet the needs of consumers and thereby provide product satisfaction.

Quality is an ambiguous term that is easily misunderstood depending on the context in which it is used. In everyday speech, its synonyms range from luxury and merit to excellence and value. In terms of academic literature, the concept of quality varies with the group using them. Each group has a different analytical framework and its own terminology. Marketing people, engineers, and manufacturers have different interpretations of quality: user-based, product-based, and manufacturing-based approaches. This frequently results in conflicts and serious breakdowns in communication. To overcome this problem, a broader perspective is required on quality within these three approaches. All the principal approaches to quality are vague and imprecise when it comes to describing the basic elements of product quality (Garvin, 1988).

Garvin identifies eight categories of quality:

- **Performance**: the primary operating characteristics of a product, e.g., acceleration, handling, cruising speed, and comfort for an automobile; sound, picture clarity, color for a television set, etc.
- **Features**: those secondary characteristics that supplement the product's basic functioning, e.g., different fabric cycles on a washing machine, automatic tuners, stereo sound on a television set, etc.
- **Conformance**: the degree to which a product's design and operating characteristics meet pre-established standards.
- **Durability**: the amount of use one gets from a product before it physically deteriorates or needs replacement.
- **Reliability:** the probability of a product's malfunctioning or failing within a specific period.
- **Serviceability**: the speed, courtesy, competence, and ease of repair.
- **Perceived quality**: composed of indirect perceptions inferred from various aspects of the product. Image and reputation of the product, for instance.
- **Aesthetics**: how a product looks, feels, sounds, tastes, or smells.

These categories represent diverse concepts: measurable product attributes, individual preferences, objectivity, time, fashion, inherent characteristics of goods, attributed characteristics, etc. The diversity of these concepts helps to explain the relationship between the different approaches and the categories of quality: the product-based approach on performance, features, and durability; the user-based approach, on aesthetics and perceived quality; and the manufacturing-based approach, on conformance and reliability. Recently, Noor et al. (2019) used Garvin's model to investigate customer satisfaction in purchasing online apparel and that performance, reliability, conformance, and aesthetics are the factors that most influence customer satisfaction in an online environment.

The rapid growth of a very competitive market requires quality in all aspects of the company's operations, with things being done the right first time and defects and waste eradicated, as much as possible, from operations. This kind of approach is known as **Total Quality Management (TQM)**. The goal of TQM is to base product development on customer needs.

Customer satisfaction is the company's highest priority and is obtained by providing a high-quality product and continuously improving the quality of the product to maintain a high level of customer satisfaction (Goetsch and Davis, 2015; Erhorn and Stark, 1994). Thus, it is no exaggeration to say that a thorough and accurate understanding of customer and market demands is the key to successful new product development.

Customer needs and product specifications are useful for guiding the conceptual phase of product design. However, during the later activities of the product development phases, teams often have difficulty linking needs and requirements to the specific design issues they face (Ulrich and Eppinger, 2019). According to the authors, for this reason, *Design for X* (DFX) methodologies is usually practiced by teams. In the Design for X (Design for Excellence) the "X" may correspond to one of the dozens of quality criteria such as reliability, robustness, serviceability, environmental impact, or manufacturability. Typical considerations in Design for X are cost, quality, reliability, and recyclability. Examples of these methodologies are Design for Manufacturability (DFM), Design for Assembly (DFA), Design for Testing/Testability (DFT), Design for Safety, and Design for Automation.

There are some methods that try to anticipate potential problems in manufacturing to the product design stage like, for example, Functional Cost Analysis, Failure Model and Effect Analysis, Functional Trees, Taguchi Method, Quality Function Deployment, Kansei Engineering, and so on. The unique methods which are based firmly on an assessment of user needs are **Quality Function Deployment (QFD)** and **Kansei Engineering**. They will be discussed in the next two sub-sections.

6.2   QUALITY FUNCTION DEPLOYMENT

Quality Function Deployment (QFD) could be defined as a product or service development process based on inter-functional teams (marketing, manufacturing, and engineering) who use a series of matrices, which look like "houses", to deploy customer input throughout design, manufacturing, and service delivery (American Society of Quality – ASQ, 2019; Maritan, 2015; Griffin and Hauser, 1993). According to the authors, QFD uses perceptions of customer needs as a lens by which to understand new product characteristics and service policies affecting customer preference, satisfaction, and, ultimately, sales. Akao (2004) defines QFD as a method to develop quality aiming at customer satisfaction, translating their demands into design objectives, and ensuring quality in the manufacturing phase. For the American Society of Quality (ASQ, 2020), QFD is a method focused on listening carefully to the customer's voice and then responding effectively to those needs and expectations.

ATTENTION

The main goal of QFD is to ensure that customer satisfaction and consumer needs are its inputs. In truth, QFD is a method that tries to translate the "voice of the user" into product requirements. In other words, it translates the user's demands into design targets and major assurance points to be used throughout the production stage.

QFD is a way to ensure design quality while the product is still in the design stage. It is particularly suitable for complex products or processes and should not be used as an isolated process in the sector of a company or a supplier. However, it applies well as a teaching design exercise in a design or engineering school.

Quality function deployment has been broadly used, in the last decade, by hundreds of companies worldwide. QFD was originated at Mitsubishi's Kobe Shipyards – Japan, in 1972 and was subsequently brought to the United States in the middle of the '80s for initial application at Ford and Xerox. QFD is now at a mature stage of implementation and can sustain the claim that it is an effective tool for systematic capture of consumer needs and addressing those needs in a structured manner within multi-functional product development teams. Several different kinds of industries have successfully applied QFD, notably for automobiles, aerospace, copiers, defense, consumer goods, electronics, textiles, computers (main-frame, mid-range, workstation, and personal), and software.

Successful accounts of using QFD are reported by several authors, including Maritan (2015), Terminko (1997), Zairi (1993), Pugh (1991), Sullivan (1986), and King (1989). Sullivan (*op. cit.*) reported that in 1979, two years after the Japanese automobile company, Toyota, had launched a new van, the use of QFD enables the company to obtain a reduction of 20% in their start-up costs; a further reduction of 38% in 1984 and a cumulative 61% reduction in 1984. During this period, affirms the author, the product development cycle was reduced by one third with a corresponding improvement in quality because of a reduction in the number of engineering changes.

QFD uses a visual data-presentation format carried out by a series of transition matrices which have a similar structure (in the form of houses). Although QFD could basically use four "houses", this number could vary depending on the properties and complexity of the product and the level of detail required. The four main linked houses conveying the user's voice through to manufacturing are named: **House of Quality** (HOQ), **Parts Deployment**, **Process Planning**, and **Production Planning**. The process of deploying the Houses of Quality in the QFD is shown as follows (Figure 6.1). From this figure, it is important to observe that the "hows" – the roof (House I, Figure 6.1) – of the HOQ (engineering characteristics) are transformed into the "what" of House II. In its turn, the "hows" of the Parts Deployment' house (parts characteristics, House II) are converted into the "whats" of the next house (House III), and so forth.

6.2.1 THE HOUSE OF QUALITY

QFD's **House of Quality** links the user's need to the desired and specific characteristics of the product. The **House of Quality** is a way of summarizing basic data in a

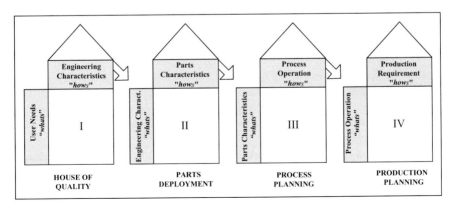

FIGURE 6.1   Components of the QFD House of Quality. Source: The author.

useful way for engineers. It represents the user's voice to the design and marketing team and is a method for discovering strategic opportunities for managers. In fact, the house encourages all of these groups to work together to understand each other's priorities and goals.

An overview of the **House of Quality** (House I) that constitutes the QFD process is shown in Figure 7.4, in the next chapter (adapted from Maritan, 2015; Akao, 1990; Hauser and Clausing, 1988; Menon et al., 1994; Pugh, 1991; Sullivan, 1986). The other Houses (II, III, and IV) are not subsequently analyzed as they relate to the manufacturing process and not to product design.

Perhaps the most important step when designing a **House of Quality** is capturing the user's needs. Having an accurate "voice" from the user is critical to the success of QFD.

6.2.2   PARTS DEPLOYMENT

The second house of QFD (see II in Figure 6.1) links **Engineering Characteristics** to actions to be taken to define **Part Characteristics**. An Engineering Characteristic (e.g., "bending system in the aluminum tubbing") – the "how" from the analyzed HOQ – can become the Part Characteristics – the "what" – of the Parts Deployment house. These Parts Characteristics can "include flexible junctions in the middle part at the ends of the tubing", for instance.

6.2.3   PROCESS PLANNING

The third matrix of QFD is the **Process Planning** (see III in Figure 6.1). It links action to implementation decisions such as manufacturing process operation. Once more, the "how" of the previous matrix – Part Deployment house – becomes the "what" of this matrix. For example, include "flexible junctions in the middle part and at the end of the tubing" ("how") of Part Deployment house will be allocated in the vertical column ("what") of the Process Planning house. The "what" of this matrix can deploy important process operations, like "make a hole in each tip of the tubing and insert a screw in each": the "how" of this matrix.

6.2.4   PRODUCTION PLANNING

Finally, the fourth house (see IV in Figure 6.1) links the planning process operation to **Production Planning** with detailed operation requirements. The key process operations, like "make a hole in each tip of the tubing and insert a screw", become the "whats", and production requirements – operator training, diameters, and other dimensions – become the "hows".

The success of QFD is strongly linked to the organization of the team. All difficulties in maintaining communication and conflicting objectives among team members must be overcome. One other important issue to be taken into account when applying QFD is the adequacy of the support tools. The product development process is frequently so detailed and complicated that no individual can comprehend it all. The implementation of QFD can falter through the lack of suitable tools – an applied computer technology, for instance, to guide the team through the maze of information.

6.2.5   THE PHASES OF THE HOUSE OF QUALITY

The **House of Quality** (HOQ) of QFD links customer needs to the desired and specific product characteristics. The HOQ is a way to summarize basic data in a usable form for the designers and engineers. It represents the customer's voice for the marketing team and is a method to discover strategic opportunities for managers. Indeed, the house encourages all of these groups to work together to understand one another's priorities and goals. Figure 6.2 shows the components of the QFD House of Quality and their respective phases. Please, pay attention that all the components are numbered appropriately.

ATTENTION

The design of the House of Quality consists of eight distinct phases:

- First phase – Identifying User Needs (The "whats"),
- Second phase – Attributing relative-importance weights to User Needs,
- Third phase – Establishing product characteristics (The "hows"),
- Fourth Phase – Establishing a relationship between different engineering characteristics,
- Fifth Phase – Designing the Relationship Matrix,
- Sixth Phase – Identifying consumer perceptions and service complaints,
- Seventh Phase – Assessing competitors, and
- Eighth Phase – Defining technical difficulty and objective target value.

The eight phases of the **House of Quality** are described below:

First Phase – Identifying User Needs (The "Whats")

The process starts by capturing what the **User Needs** (user requirements) in the product and establishing a relative prioritization. This will generate a "What list", the basis of the **User requirement** component (Component 1, in Figure 6.2).

User Needs are composed of the consumers' own verbalizations and are obtained from market research and competitor analysis. They are used to describe products and product characteristics. A QFD matrix for the design of a wheelchair is illustrated in Figure 7.4, next chapter. It shows the house's basic concept. A typical application has 30 to 100 **User Needs**, such as: "reduce the weight of wheelchairs", "produce foldable wheelchairs", "allow easy traverse of difficult terrain", "easy to remove wheels", etc. Some may include demands of regulators ("safe in a side collision"), needs of retailers ("easy to display"), requirements of vendors ("satisfy assembly and service organizations"), and so forth. The accuracy and quality of this first phase are crucial

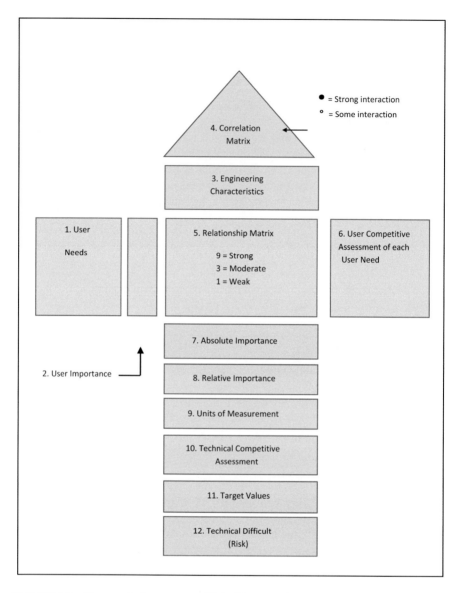

FIGURE 6.2 The translation process of linked houses of QFD. Source: The author.

to what follows. It is also the most difficult phase because it requires the procurement of real user needs and not what the team thinks that user needs.

Second Phase – Defining Relative-Importance Weights to User Needs

Some requirements have different levels of importance for the user, and to satisfy them, designers have to trade off one benefit against another; for this it should assign weights to the needs that users consider most important. Statistical techniques can be used to allow users to state their preferences with respect to existing and hypothetical products. **Weighting**, representing prioritization of User Needs, is displayed in the HOQ, after the column of User Needs (Component 2, in Figure 6.2).

Third Phase – Establishing Engineering Characteristics (The "Hows")

Users' needs are typically subjective expressions helpful in developing an understanding of what the users want. However, they offer little guidance about how to design and engineer the product (Ulrich and Eppinger, 2019). Phase 3 of the design of HOQ corresponds to describing the **User Needs** in the language of the designer/engineer. The team identifies those measurable aspects of the product or service, which, if modified, would affect users' perceptions generating a "How list" of design attributes. Engineering/Product characteristics or requirements depend on the use of which the product is put. Along the top of the matrix is a list of those technical characteristics that affect one or more user attributes. These characteristics – named **Engineering Characteristics/Product Characteristics** (ECs, Component 3 in Figure 6.2) – will be developed as the basis for subsequent product design and process development and must be described in measurable terms. Such characteristics will be deployed through design, manufacturing, assembly, and technical assistance in such a form that a product's final performance meets user needs.

Fourth Phase – Defining the Correlation Matrix

The HOQ's roof is the **Correlation Matrix** (Component 4, Figure 6.2), which specifies the relationship among the Engineering Characteristics. It helps the design/engineering team to specify the several engineering features that have to be improved collaterally (presented as either "strong interaction" or "some interaction"). Its function is to help the design/engineering team specify the various engineering features that need to be improved from a cross-verification. This cross-verification permits identification of critical information when designers/engineers need to balance the trade-offs in terms of user benefits.

Fifth, Sixth, and Seventh Phases – Designing the Relationship Matrix and Defining Absolute and Relative Importances

The **Relationship Matrix** (Component 5, Figure 6.2), the body of the HOQ, is a relationship matrix that displays judgments (or experiments) indicating which design attributes or **Engineering Characteristics** items affect **User Needs** by how much. The degree of correlation will be defined using appropriate numbers (9 = strong, 3 = moderate, and 1 = weak) to determine the strength of each correlation. This evaluation will be established by the project team, in a consensual way, based on experience, user responses, statistical studies, or controlled experiments. The **Absolute** and **Relative Importance** will be given for each **Engineering Characteristics** (Components 6 and 7, in Figure 6.2). See details on how to attribute absolute and relative importance in sub-chapter 7.2.4 (Applying Quality Function Deployment to Product Development).

Eighth Phase – Consumer Evaluation (Identifying Consumer Perceptions and Service Complaints)

This phase aims to identify consumer perceptions and complaints about the services provided. It is characterized by identifying the degree of competitiveness, using a user preference chart, to obtain user perceptions between in-house and competitor products and service complaints related to each **User Needs** (Component 8, Figure 6.2). Ideally, these evaluations are based on scientific surveys of customers carried out by marketing teams. This section of the HOQ enables direct assessment of the proposed specification and determines the potential positioning of the in-house product against the competition. This procedure identifies strong and weak points of the product and gives the opportunity for improvement.

Ninth Phase – Units of Measurement (Assessing Competitors)

In this phase, **Units of Measurement** are used to evaluate competitors and are given to each Engineering Characteristics (Component 9, Figure 6.2). The aim of this phase is to carry out a comparison with the competitor's specifications for each of the product's Engineering Characteristics.

Tenth, Eleventh, and Twelfth Phases – Competitive Technical Evaluation (Defining Technical Difficulty, Objective Target Values, and Technical Difficulties)

After the team has identified the user requirements through measurements and linked them to the Engineering Characteristics, the **Competitive Technical Evaluation**, the **Target Values**, and **Technical Difficulties** that the company will pursue for the new project should be included at the bottom of the House of Quality.

In the **Competitive Technical Evaluation** each value identified in the Engineering Characteristic is presented for each competitor (component 10, Figure 6.2). In the **Target Values** it is sought to identify measures (values) for each Engineering Characteristic (component 11, Figure 6.2). Finally, at the base of the House of Quality, the **Technical Difficulties** (component 12, Figure 6.2) are identified, based on risks associated with changes in design attributes, product characteristics, and engineering requirements.

The **House of Quality** of **Quality Function Deployment** is now completed. It is important to observe that the detailed structure of the matrices can vary depending on the application and that it is a flexible tool allowing changes to be made to adapt to the design.

The process of QFD continues after finishing the **House of Quality**. As presented earlier in Figure 6.1), other linked houses conveying the user's voice through to manufacturing will be designed. These houses have the same structure as the HOQ, and the "hows" of one stage becomes the "whats" of the next.

6.3   *KANSEI* ENGINEERING/AFFECTIVE ENGINEERING

No discussion of human-centered design methodologies would be complete without some reference to *Kansei Engineering* **(KE)/Affective Engineering**. It is a consumer-oriented product development technology that aims to transform customers' perceptions, feeling, and mental images into tangible products (Nagamashi, 1995, 2016b). When a consumer wants to buy a new product, he or she expresses the wish with words such as "gorgeous, beautiful and strong at an inexpensive price". KE is able to interpret and transfer the psychological implications of these words to the details of the design of the product.

According to Nagamashi (2016b), *kansei* is a Japanese word that has no exact English word to describe it. The closest translation for *kansei* can be feeling, sensibility, and emotion (Lokman, 2010; Yoshikawa, 2000, Nagamachi, 2002), but none of them is said to appropriately denote it. For this reason, *kansei* is used in its Japanese form.

Schütte et al. (2004) stated that **Kansei Engineering** is a proactive product development methodology, which translates the impressions, feelings, and demands of users about existing products or concepts into design solutions and "concrete" design parameters.

According to Nagamashi (2016b), in addition to helping the customer to select a product that fits his or her feeling, *Kansei* Engineering also provides the designers with a tool to link consumers' feelings and design. To obtain relations between *Kansei* and design details, several analyzes must be made to determine which types of external appearances and functions produce what kind of feelings.

Using the Kansei user experience as a principle, a series of mathematical correlations is performed relating Kansei to the physical characteristics of the product, aiming to improve human well-being based on the analysis of its physical and psychological aspects that contribute to satisfaction (Lokman, 2010).

Kansei Engineering was applied to several products, including fashion design, car doors, interior design of cars and homes, furniture, and office chairs (Nagamashi, 2017, 2016a, b; 2002). Lee et al. (2002) discuss how to apply and evaluate Kansei in product design from abstract images, dynamic manipulation of 3D objects and how shapes are perceived by participants in an experiment.

TIP

Want to know more about *Kansei* Engineering? Read:

Mitsuo Nagamashi (2017). *Innovations of Kansei Engineering*, CRC Press. ISBN-13: 978-1138440609.
Simon T. W. Schütte, Jörgen Eklund, Jan R.C. Axelsson and Mitsuo Nagamachi (2007), Concepts, Methods and Tools in *Kansei Engineering*. *Theoretical Issues in Ergonomics Science*, 5:3, 214–231, DOI: 10.1080/1463922021000049980.

Another methodology proposed by the Japanese school of product evaluation is **Kawaii Engineering** (Ohkura, 2019). According to the author, "Kawaii" is a Japanese word that denotes "cute", "adorable", or "charming", although it does not have exactly the same meaning as these adjectives. The author presents an engineering methodology for systematic measurement of the affective perception of *kawaii*, using virtual reality and biological signals, and discusses the effectiveness of Kawaii Engineering to design industrial products and services. Kawaii Engineering contributes to attracting people's sympathy for a so-called "cute design", which reduces fear and makes boring information more acceptable and attractive.

6.4 PRODUCT USABILITY

There are many examples of bad design that we find in our daily lives. Bad design means bad usability. According to the concept of good design presented in sub-chapter 2.2, there is no product with good design and poor usability. Good usability means good design. These concepts are intrinsically related. Thus, we will define and analyze usability and user experience in this and the following sub-chapter.

Usability is defined by the International Standards Organization as "the extent to which a product can be used by certain users to achieve specific goals with **effectiveness, efficiency** and **satisfaction** in a certain context of use" (ISO 9241-11: 2018 (en), 2019c).

Jordan (1998a) analyzes these concepts defining in detail effectiveness, efficiency, and satisfaction:

EFFECTIVENESS

- This refers to the dimension by which an objective or task is achieved. Effectiveness measures the relationship between the results obtained and the intended objectives. That is, to be effective is to achieve a given objective.
- In some cases, the distinction between a task to be performed or not can be characterized simply as the success or failure in carrying out the task. For example, a drill is effective if it can drill a hole in the wall.

EFFICIENCY

This refers to the amount of effort required to achieve an objective. The less effort, the greater the efficiency. The effort can be measured, for example, in terms of time to complete a task, the number of errors that the user makes before the task is completed, or the energy expenditure to complete a task. For example, a drill is efficient if it can drill a hole in the wall without causing too much vibration or physical effort to the user.

SATISFACTION

- This refers to the level of comfort that the user feels when using a product and how acceptable the product is for the user to reach his goals. It can also be defined as how well a product or service meets users' expectations.
- This aspect is more subjective than effectiveness and efficiency and can also be more difficult to measure.

ATTENTION

There are several definitions of usability in the literature. Here are some of them:

- "Usability is a quality attribute that assesses the ease of use of user interfaces. The word "usability" also refers to methods to improve ease of use during the design process" (Nielsen Norman Group 2020).
- "Design principles that – when followed – provide answers to users, making using the device easier" (Norman 2013).
- "A set of four factors gathered in one device: 1) the ability to be used successfully; 2) ease of use; 3) user's ability to learn to use the device in a simple and quick way; 4) Provide visual user satisfaction" (Rubin and Chisnel 2008).
- Ability, in human functional terms, for a system to be used easily and efficiently" (Shackel 2009).
- "A set of properties of an interface that brings together the following components: 1) Easy to learn; 2) Efficiency; 3) Storage capacity; 4) Low error rate; 5) Satisfaction and pleasure in use" (Nielsen 1993).

There are countless examples on the internet about bad design and consequent bad user interface, with products and systems that result in poor usability. For instance, chairs of inadequate size for larger or smaller users; keyboards with an inappropriate design that does not allow the user to type with the handles in a neutral position, without lateral deviation; information that does not allow for quick and effective decision making; a poor positioning of controls and triggers that forces the assumption of inappropriate postures, etc. The user is often blamed for the flaws in these interfaces, in fact, it is very easy to blame the user, but it is not always true that it is his or her fault.

Designing a good product is not easy. Product development is a risky business because it involves, at a high cost, several sectors of the company. The design process can reduce the risk and/or cost of product failures.

ATTENTION

In general, usability requires that the product or system achieve these goals:

- Be **efficient** in its use.
- Be **easy to learn**.
- Be **easy to remember**.
- Show **few errors** in use.
- Be **pleasant**.
- Ensure user **satisfaction**.

To achieve the goals proposed above, products and systems need to ensure a good interface.

6.4.1 HUMAN–PRODUCT INTERFACE

The user interface is the boundary between the user and the functional part of a system. The user experience occurs as a result of this interface.

We present a model of the **Human–Product Interface** (Figure 6.3), based on Kroemer and Grandjean (2004) and Moraes and Mont'Alvão (2010). We can see that in this model, we have the representation of the **Product** on the one hand and the **Human Being** on the other. Here, the term "product" is being described in a generic way (a pen, a computer, a lathe, or a panel console for a nuclear power plant).

The smart products currently on the market have changed the traditional interface of the human-task-machine system. Smart products combine physical and software aspects. For example: in the new smart refrigerators, the user interacts with the physical product (the refrigerator) and the digital representations (the screen) that provide a new interface; new television sets allow the interface through gestural or voice commands. Soares et al. (2021) point out that smart products are able to interact with the user using sensors and input and output data, and have the ability to adjust to the environment, being sensitive to the situation and context of use.

Thus, according to the model presented, the **Product** interacts with the **Human** through **Information**, which will be perceived by the user, and consequently transformed into **Actions**. The product information (which can be visual, auditory, or haptic) is sent to the human through displays (visual), speakers/sound systems (auditory), or vibrations (haptics) produced by the **Information Devices** (displays, speakers,

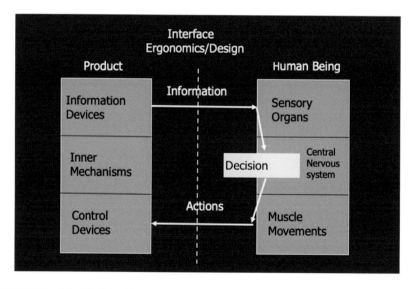

FIGURE 6.3   Model of the Human–Product Interface. Source: The author.

haptic screens, etc.). It is worth mentioning the concept of haptics used in this book. **Haptic** is understood as the vibrating or tactile characteristic of some equipment, such as the vibrating or touch screen feature of a smartphone. The information is received by the human through some **Sensory Organs** (sight, hearing, and touch).

Then, the Information is processed by the c**entral nervous system**, which is responsible for triggering **Decision** making, such as the movement of a segment of the body involving its joints and muscles. These **Decisions** will allow the human to exert **Actions** on the product through **Muscle movements**: with the hands (palmar action), fingers (touch action), body (gesture command), or even speech (voice command). Such Actions on the product Control Device will activate **Internal Mechanisms** responsible for changes in the system. The changes will be presented again by the Information Devices. This means that the **Human–Product System** circle will be closed and will be continuously changed with new actions.

An example to illustrate this model of Human–Product Interface is the use of a smartphone by the user. When the user receives a call, it is perceived when his sensory organs (hearing, sight, or touch) perceive the information transmitted by the information device (sound, image on the display, or vibration of the device). The user will decode the information from the central nervous system and make a decision that, in this case, will correspond to the task of answering or ignoring the call. If the user decides to attend, he or she will activate muscle movements which, in turn, will perform actions on the product control devices. Actions must be expressed in verbs, in this case, "touch", that is, "with your fingers touch the smartphone screen". The control device will activate the product's internal mechanisms responsible for its operation. New information will be presented on the information devices from the action taken, and this interface cycle will always be fed back.

The interface between the product and the human occurs through information and action. Thus, the role of the designer, with the support of ergonomics, is to design and optimize the interface, that is, to project product information compatible with the user's physiological characteristics, so that he or she can perform appropriate actions considering his/her physical, cognitive conditions and repertoire of knowledge. The product information can be, for example, the font size of the display, figure x background ratio, sound intensity, vibration characteristics, etc., that will be related to sensory/human receptors. Actions on the product can be touching, pressing, suspending, turning, pulling, pushing, etc., that are related to muscle movements of the hands or feet.

This model allows us to analyze in detail the information, the required actions, and the design requirements of a product with a view to its optimization (e.g., reading from a source on display from a certain distance, anthropometric data necessary for body parts to perform a given action, etc.). This requires knowledge of the human body and cognitive aspects of the user provided by ergonomics. However, it is necessary to rethink this model with the advent of new technologies provided by smart products.

The smart products currently on the market have changed the traditional form of the user interface with the system. Smart products combine physical and software aspects. These products can also interact with each other, making an interconnection between everyday objects and the internet, connecting more objects than people. This type of interaction is called the **Internet of Things**, enabling new experiences in the human-product field.

However, direct user–product interaction is still the most common, such as new smart refrigerators. The user interacts with the physical product (refrigerator) and digital representations (screen) that provide a new interface; new television sets that allow the interface through gestural or voice commands. Faced with all this technological revolution, designers of the future also need to think about products that can aggregate extended realities (Augmented Reality/AR, Virtual Reality/VR), Artificial Intelligence (AI), and commands through the brain–machine interface. The human–product interface should be increasingly natural and intuitive.

According to Leventhal and Barnes (2007), the factors that influence the success of an interface are:

- User characteristics.
- The type of task to be performed.
- Hardware limitations.
- Social and cultural limitations.

For this, the authors continue, the following characteristics that interfere with usability must be considered:

- The user experience with the interface.
- Domains of knowledge and cultural aspects of the user.
- Any eventual disability.
- Age and sex, in this case, in terms of the users' anthropometric dimensions.

Therefore, such factors and characteristics should remain in the human-product relationship regardless of the advancement of technology.

ACTIVITY

- Think of an activity that you perform using a product.
- Identify the product with which you interact.
- Define your tasks and interactions with the product (try to represent them using a graphic format).
- Introduce the human–product interface model.
- Reflect if the product contains the usability objectives previously presented (see previous box "Attention" that presents the usability requires of a product).

6.4.2 USABILITY MODELS

Soares et al. (2021) state that a usability evaluation focuses on understanding how people use a product with respect to the interaction between the user, the task, and the product. To carry out a usability evaluation, some authors have proposed evaluation models that will be presented below. The Shackel (2009), Eason (2007), and

Nielsen (1993) models are originally proposed for software environments. Jordan's model (1998a) is specific to products, while Leventhal and Barnes (2007) present a hybrid model that can be used for both product and software analysis.

(i) Shackel's usability model (2009)

The Shackel model has four main characteristics or dimensions of usability:

- **Effectiveness** – evaluates the level of performance required to complete a task, compared to the percentage required by standard users in a given usage environment.
- **Learning** – the ease of learning measured from the time after the installation of the system and relearned for intermittent users.
- **Flexibility** – allow adaptations beyond those originally specified for the system.
- **Attitude** – the system must be used within acceptable human costs. Human costs are understood as tiredness, discomfort, frustration, and personal effort.

(ii) Nielsen's usability model (1993)

In the Nielsen model, usability is part of a broader approach regarding system acceptability. According to the author, the acceptability of the system is defined as the sum of social acceptability, practical acceptability, and usability.

- **Social acceptability** – involves acceptance of the system by the population of users for whom it is intended.
- **Practical acceptability** – involves reliability, compatibility, costs, and usefulness (usability would be included in this dimension of practical acceptability).
- **Usability** – involves factors such as Ease of learning, Efficiency in use, Ease of remembering, Few errors, and Pleasantness.

(iii) Eason's usability model (2007)

The usability model of a system presented by Eason is based on the characteristics of the user, the interaction of the user with the system and the characteristics of the task. It is defined in the context of the purpose for which it is being used and who the users are.

Thus, Eason points out three determinants for usability:

- **Ease of use** – means the effort that is required to operate the system after it is understood and mastered by the user.
- **Ease of learning** – means the effort that is required to understand and operate a system that is unfamiliar.
- **Suitability for the task** – means how much information and functions a system provides in order to suit the user's needs for a given task. To what extent the information and functions of a system correspond to the user's needs to perform a specific task.

Eason indicates that ease of use, ease of learning, and compatibility of interface tasks are significant determinants of usability. These notions are different. A system that is difficult to learn initially because it has many commands to memorize can be easy to use after the user has memorized the commands.

For a better understanding of the Eason Model, it is important to analyze the characteristics of the task and the user.

- **Task characteristics** – The task is what you do with the device and can be more or less independent of the specific system. It involves **Frequency** (number of times the task is performed by the user) and **Openness** (number of options that the interface offers to perform the task).
- **User characteristics** – involves **Knowledge** (means the knowledge that the user applies to the task), **Motivation** (if users have a high degree of motivation, they will more easily overcome physically and cognitively more demanding tasks), and **Criterion** (refers to the ability to choose to use (or not use) some part of the system with which it is interacting).
- Other user characteristics: **Learning style**, **Ability to solve problems**, **Age** and how it influences the interaction, **Physical characteristics,** and **Personal skills**.

Falcão and Soares (2013) analyzed the models that had originally been developed to analyze software and websites proposed by Shackel (2009), Eason (2007), and Nielsen (1993) and concluded that they could be perfectly adapted for use in consumer products. The authors presented a taxonomy of the three models based on Shackel (2009), Eason (2007), and Nielsen (1993) (Table 6.1).

(iv) Usability model by Leventhal and Barnes (2007)

Laura Leventhal and Julie Barnes present a hybrid model, based on Shackel, Nielsen, and Eason. In this model, they have three dimensions regarding the task, the user, and the user interface (Figure 6.4), presented below:

- **Task variables**: **Frequency** (tasks that are performed frequently are likely to include well-learned sequences of user actions), **Rigidity** (is defined from the number of paths to be taken through the task and the number of options and limitations that are available along the paths) and **Situational constraints** (how much the system imposes restrictions regarding, for example, security, interaction environment, individual or collaborative use, limitations in verbal and/or visual instructions, etc.).
- **User variables**: **Expertise** (the user has the mastery of the task and the context of use superior to occasional and novice users), **Novices** (the user has a very limited mental model of the task and its execution), **Occasional** (is among a specialist and a novice, may have a good abstract idea of how to perform a task but may have overlooked the precise details); **User motivation** (how much the level of motivation influences the final result of the system's usability).

TABLE 6.1

Taxonomy of the usability models of Eason, Shackel and Nielsen (Falcão and Soares, 2013)

Model	Dimensions		Definition
Eason's Model (2007)	Task	Frequency	Number of times a task is performed by the user.
		Openness	Extent to which a task is modifiable.
	User	Knowledge	The knowledge that the user applies to the task and whether it is appropriate or not.
		Motivation	What determines that the user fulfills his task.
		Discretion	The user's ability not to choose to use any part of the system.
	System	Easy of learning	The effort required to understand and operate an unfamiliar system.
		Easy to use	The effort required to operate a system once it is understood and mastered by the user.
		Task match	The extent to which each information and function that a system provides corresponds to the user's needs for a given task.
Shackel's Model (2009)	Effectiveness		It is described by the results of the interaction in terms of speed and errors made.
	Learnability		Corresponds to the time between the training of users and the frequency of use, including the time of relearning.
	Flexibility		Corresponds to the adaptation of some tasks and/or environments in addition to what was specified first.
	Attitude		Corresponds to acceptable levels of the human cost in terms of tiredness, discomfort, frustration, and personal effort.
Nielsen's Model (1993)	Easy to learn		The system must be easy to learn so that the user, even without having experience, can quickly begin to obtain satisfactory results from work performed.
	Efficient to use		It is directly related to the productivity of the system so that once the user has learned the system, high productivity is possible.
	Easy to remember		The system should be easy to remember so that the occasional user does not have to learn everything about the system again after some time without using it.
	Few errors		The system must have a low error rate so that users make fewer mistakes while using the system, and as soon as errors are made, they can be corrected simply and quickly. In addition, catastrophic errors should not occur.
	Subjectively pleasing (Satisfaction)		The system should allow pleasant interaction so that users are subjectively satisfied when using it.

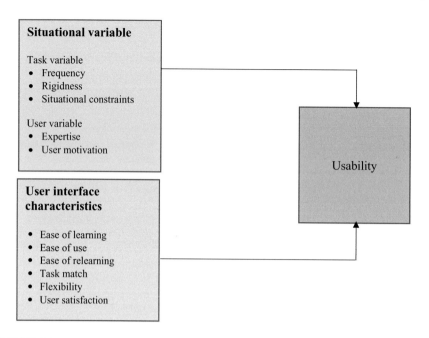

FIGURE 6.4   Usability model proposed by Leventhal and Barnes (2008), adapted by Falcão and Soares (2013).

- **User interface characteristics**: **Ease of learning** (how easy is the interface for new users to learn to use the product or system), **Ease of use** (how easy is the operation of the interface for users), **Ease of relearning** (how easy it is to relearn an interface after it has been learned once, but not used recently), **task matching** (it's a good relationship between the interface and the task in order to allow users to develop a better understanding to perform the task), **Flexibility** (refers to the interface's ability to support unforeseen usage patterns) and **User Satisfaction** (can be defined as the user's response to the assessment of the perceived discrepancy between the expectations existing before the purchase and the actual product performance perceived after consumption).

Leventhal and Barnes (2007) state that a usability model can be applied to several phases of the analysis of a product's user interface, such as:

- Assess the usability of an existing system and the user interface.
- Use customer demands as beacons to analyze usability characteristics, along with information from situational variables to conduct a project and define usable interfaces.
- Monitor and evaluate usability during the development of the interface.
- Analyze the final product.

ACTIVITY

- Choose a consumer product to evaluate the user interface using the usability model by Leventhal and Barnes (2007).
- Describe the tasks of the user who handles the device. Explain how user input results in device actions. Make a drawing to accompany the text description.
- Identify the product measurements for each of the user characteristics, characteristics of the user interface, and characteristics of the task. You must not take measurements. For each measurement, just describe what it is, how you would make the measurement, and why the measurement is appropriate for that characteristic.

(v) Jordan's usability model (1998a)

Jordan's usability model applies more to consumer product analysis. The following components are part of this model:

- **Guessability** – It is the measure of cost to the user when using a product and performing a new task for the first time.
- **Learnability** – Refers to the cost for the user to reach a competent level of performance in a specific task assigned to a product, having already performed that task before.
- **Experienced user performance** – Refers to the relatively unchanged performance of someone who has used a product many times to accomplish a particular task.
- **System potential** – Represents the maximum level of performance that would be theoretically possible to complete a task with a product.
- **Re-usability** – Refers to a possible reduction in performance after the user has not used the product for a relatively long period.

Jordan (1998a) presents the following Usability Principles applied to product evaluation:

- **Consistency** – similar tasks must be performed in similar ways, e.g., driving a vehicle's pedals must be similarly designed regardless of make or model.
- **Compatibility** – design a product so that its method of operation is compatible with users' expectations based on their knowledge of other types of products, e.g., use the green or red button for permitted or prohibited tasks, based on population stereotypes.
- **Consideration of user resources** – when designing a product, consider the physical and physiological aspects of users and their dimensional data and physiological limits of strength, precision, etc., e.g., when designing a TV

remote control, consider the catch, sight, and cognitive aspects to understand the messages.

- **Feedback** – is the response of a product or system to any actions that the user has taken, e.g., the sound emitted by the computer as a response to the performance of an activity that is not allowed.
- **Prevention and recovery of errors** – if the user makes an error, it can be predicted and corrected, quickly and easily, without prejudice to the task being performed, e.g., impediment or quick correction of the action of typing a letter in a field where numbers are only allowed in a spreadsheet.
- **User control** – allowing the user to have full control over the actions performed on the product, e.g., easy-to-use settings on an office chair.
- **Visual clarity** – allow the user to read and understand information clearly, quickly, and unambiguously with color codes compatible with the action to be performed, e.g., information on a TV screen presented by the remote control.
- **Prioritization of functionality and information** – design controls and displays arranged in such a way that those in constant use are more accessible to users, e.g., a camera whose controls and information are the most used are easier to access and view.
- **Appropriate transfer of technology** – use technology developed in another context to improve usability, e.g., information projected on the windshield of an automobile with more advanced technology is head-up displays originally designed for military aircraft.
- **Explicitness** – are tips provided by products and systems that explain their functionality and the method of operation, e.g., an anti-panic bar on a fire door that allows quick decision making.

The application of usability models refers to the effective evaluation or test of the usability of a product or system. For this, the variables of interest that need to be properly measured must be defined. Metrics must also be established, and measurements defined to be interpreted by the designers, engineers, and those responsible for executing the product.

6.4.3   Research Methods and Usability Assessment

The usability survey data can be presented as:

- **Quantitative data** – numerical values are used, and the data is more accurately described.
- **Qualitative data** – data that reflect users' feelings and thoughts based on subjective opinions.

Leventhal and Barnes (2007) recommend using quantitative or qualitative data to assess usability metrics, as shown in Table 6.2, below:

TABLE 6.2

Data examples and corresponding metrics presented by Leventhal and Barnes (2007)

Data	Metrics
Ease of learning, ease of use, and ease of relearning	Precision performance through speed and error rate.
Task matching and the flexibility of the user interface to support new tasks	Expert opinions through quantitative measurement scales or subjective opinions.
Use experience	Assess user knowledge, level of experience, or response to situations.
Motivation	User's response to situations.
Satisfaction	Rating scales.

According to the authors, other examples of some usability measures are:

- User's self-assessment on a scale.
- User's years of training or number of courses taken.
- User's score on a comprehension test.
- Test is given after learning the system.
- Number of errors on a standard task.
- Time to complete a standard task.
- Number of standard tasks completed in a set time period.
- Number of features used.

Usability assessments, tests, and experiments are usually carried out through various subjective protocols. The tools for evaluation, tests, and experiments can be divided into **empirical** (when involving participants) and **non-empirical** (when involving experts).

ATTENTION

There are many tools used to assess usability. Here are some based on Hom (2020), Cybis et al. (2017), Leventhal and Barnes (2007), Maguire (2001), and Jordan (1998a).

EMPIRICAL METHODS

- **Private camera conversations** – This method involves participants entering a private booth and talking to a video camera about a pre-defined topic set by the investigator.
- **Co-discovery** – This method involves two participants working together to explore a product and/or to discover how particular tasks are done.

- **Focus group** – The focus group is a group of people gathered together to discuss a particular issue. It consists of a discussion leader and a number of participants. The leader will have an agenda of issues that will form the borders within which the discussion can proceed.
- **User workshops** – Involves a group of participants gathered together to discuss issues relating to a product's design and usage. Usually, users will get involved in "designing" a new product. This might mean simply listing their requirements in terms of usability and functionality.
- **Think-aloud protocols** – This method involves a participant speaking about what they are doing and thinking when using an interface. Participants may be asked to perform specific tasks with an interface, or they may simply be given the opportunity for free exploration.
- **Incident diaries** – are mini-questionnaires that are issued to users in order that they can make a note of any problems which they encounter when using a product. Typically, users might be asked to give a written description of the problem that they were having. They might then be asked how they solved it (if at all) and about how troublesome the problem was.
- **Feature checklists** – a list of a product's functionality. Users are simply asked to mark against the features that they have used.
- **Logging use** – With computer software and some other software-based products, it is possible to install automatic logging devices that keep a record of users' interactions with the product.
- **Field observation or Ethnographic study** – Involves watching users in the environment in which they would normally use a product.
- **Card sorting** – through cards, the aim is to discover the mental model that the user has about information from a given interface, categorized based on their importance and significance.
- **Questionnaires** – Correspond to lists of printed or digital questions. Including in this category SUS (System Usability Scale) and Post Study System Usability Questionnaire (PSSUQ).
- **Interviews** – Here, the investigator compiles a series of questions which are then posed directly to participants
- **Controlled experiments** – An experiment is a formally designed evaluation with comparatively tight controls and balances. The aim is to remove as much noise as possible from the data in order to isolate effects for performance with the product as cleanly as possible.

NON-EMPIRICAL METHODS

- **Task analysis** – It means breaking the task down into as many observable activities as possible. It is usually recorded through a flow chart.
- **Property checklists** – It lists a number of design properties and requires the researcher to verify that the product being evaluated for conforms to the properties on the list.

- **Heuristic evaluation** – The product is evaluated based on an expert's analysis. A specialist performs an inspection of the interface to identify any problems.
- **Cognitive walkthrough** – The expert performs the assessment from the point of view of a typical user, trying to perform a certain task in order to predict any difficulties for the task to be completed.

Ensuring good usability does not mean providing pleasure when using a product (Jordan, 2002, 1998b). Jordan states that pleasure in using the product is when emotional and hedonic benefits occur associated with using the product. The user experience is the result of the interaction between the user, the product, and the system and will be presented below.

6.5   USER EXPERIENCE (UX)

User Experience (UX) affects everyone every day. It crosses cultures, sex, age, and economic class. The **User Experience** is the result of the user's interaction with the product or system (interface), largely provided by usability. Thus, all elements that act in the relationship of the user's interaction with a product, system or service will define the **User Experience**. If usability is directed to aspects related to human performance in its interaction with products or systems, UX reflects the emotional aspects arising from this interaction. Thus, usability is directly related to the tasks performed, while UX corresponds to the experiences lived by users.

Quaresma et. al. (2021) state that experience is eminently personal, whereas products or systems are external factors to the user. Experience is found in the mind and memory of users, regardless of the emotional, intellectual, physical, or spiritual levels of individuals (Pine and Gilmore, 2020). Thus, while two people may have similar experiences, it is impossible for them to live exactly the same experiences (Quaresma et al, 2021).

ISO 9241-110: 2010 (2.15) (2019b), defines UX as "the perception and response of the person that results from the use or anticipated use of a product or service". In this way, UX is related to the feelings and emotions that the user experiences when using a product, system, or service.

INTERNET

Donald Norman coined the term **User Experience** (UX). Watch in this video your testimony about the meaning of the term.

The term "UX", Don Norman, 2016.
https://www.youtube.com/watch?time_continue=1&v=9BdtGjoIN4E&feature
    =emb_logo

6.5.1 EMOTION AND FEELING

Damásio (2015) distinguishes **emotion** from **feeling**. According to the author, **emotion** is a set of bodily reactions, automatic and unconscious, in the face of certain stimuli from the environment where we are inserted. They are chemical and neural responses based on emotional memories and arise when the brain receives an external stimulus.

The **feeling** arises when we become aware of our emotions. That is, the feeling occurs when our emotions are transferred to certain areas of our brain, where they are encoded in the form of neuronal activity. Feeling is a response to emotion. UX is associated with the emotions produced by the interaction with the product, system or service.

6.5.2 EMOTION AND EMOTIONAL DESIGN

Emotions seem to dictate people's rules of behavior on a daily basis, because decisions are made based on different moods: happiness, sadness, irritation, annoyance, or frustration. Emotions are often intertwined with a series of psychological phenomena: mood, temperament, personality, and motivation.

Shin and Wang (2015) state that emotion is generally defined as a complex feeling state that responds to stimuli, through physical and psychological changes, which can influence the individual's thinking and behavior with the product.

User interaction with the product in the real world can be a source of encouragement to trigger emotions in people's daily lives. Each user interaction with the product leads to a distinct user experience (UX). Through experiments, it is possible to investigate the result of the interactions between an individual and the components that make up the product, environment, or system at a given moment (Margolin and Buchanan, 1995).

ATTENTION

Norman (2008) defines **Emotional Design** as the characteristic that a product has to arouse emotions in end users, through its design, in order to establish connections between the product and the user. The term was created by the author to study the "emotional bonds" between the user and the product and try to answer questions such as: what makes a user choose between one product or another, even if they have similar performance.

Norman relates the emotional process to three levels of human brain processing (Komninos, 2020; Norman, 2008):

- **Visceral** – responsible for the ingrained and automatic qualities of human emotion. They are emotions entirely beyond our control and are related to quick decisions about what is good and bad, safe, and dangerous; it refers to the perceptual qualities of an object and the sensations it causes. It is what, in the eyes of the user, makes a product different from the other in terms of attitudes, beliefs, and feelings.

- **Behavioral** – refers to aspects related to the control of the action; where actions are analyzed unconsciously, and strategies are defined to analyze a situation in order to deal with situations in the most effective way and in the shortest possible time; are associated with the use and user experience with a product. This level of processing is closely linked to the product's function, performance, and usability; and
- **Reflective** – refers to conscious thinking, the learning of new concepts, and their generalization about the world. It refers to aspects related to the rationalization and meaning of a product. It is linked to perceptions related to luxury, expressiveness, status. According to the classification presented by Damásio (2015), above, this emotional process would be more related to feeling, rather than emotion.

Each of these levels of processing, says Norman, involves the user's experience with the world in a specific way. Figure 6.5, based on Komninos (2020) and Norman (2008), shows the relationship of the design x product x user with the three levels of emotional process that provokes in the user.

For Damásio (2015), emotions can be divided into two types: (1) the **Innate emotions** that are related to survival needs and (2) the **Cultural emotions** that are learned and influenced by people and the environment where the individual lives.

Some studies relate the use of products with emotion and pleasure (Jordan, 2002; Green and Jordan, 2002) and perception and emotion (Seva et al., 2011). Such studies analyze user satisfaction according to the sensations of the experience with the product and the matrix mapping of the emotions generated. According to Jordan (2002), in the context of design, pleasure is defined as "the emotional, hedonic and practical relationship of the person with the product".

Thus, the author presents four categories that relate pleasure to the use of products:

- **Physical pleasure** is related to the body and the sensory organs, such as touch, smell, taste. People enjoy the smell of a book or a new car or touch the texture of a high-end cell phone. As an example, the coffee shops that use smell as a way to attract consumers.

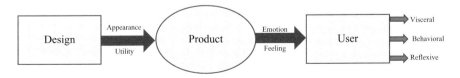

FIGURE 6.5 The relationship between Design x Product x User and the three levels of emotional processing, based on Komninos (2020) and Norman (2008).

- **Social pleasure** comes from the pleasure derived from social relationships. Some products have their use related to the status, such as a luxury car, or even a high-end smartphone.
- **Psychological pleasure** is related to the user's cognitive and emotional demands when using a product. An easy-to-use product with low cognitive demand will provide much more pleasure to the user than difficult ones, which cause discomfort or mental fatigue. Feeling challenged is also another form of psychological pleasure. Games that stimulate the brain, such as Sudoko, for example, provide the pleasurable feeling of victory.
- **Ideological pleasure** is related to the rational aspects and values of the user, it is related to social values such as right and wrong. Biodegradable and ecological products are examples of products that provide ideological pleasure to this group of users.

In order to understand the emotional aspects of user–product interaction, studies in human neurophysiology, design, cognitive sciences, and artificial intelligence have been improving in recent years. The application of neuroscience to emotional design presents itself as an ample possibility to assess users' behavior in the face of emotions caused by the product with which the user is interacting.

Esperidião-Antônio et al. (2008) state that studies using the neuroimaging technique, through an in-depth study of the limbic system (also known as the emotional brain, located below the cortex region), seek to expand knowledge about the neural bases and processes related to emotions. Therefore, the interest of these studies is to understand the relationship between emotional, cognitive, and homeostatic processes and the respective physiological responses of individuals. In other words, these analyzes make it possible to establish a relationship between emotions and brain circuits in the different situations of the individual who manipulates a product.

Helander and Khalid (2012) raise two research questions about emotion: (1) how we can measure and analyze human reactions to affective and pleasurable design, and (2) how we can evaluate the characteristics of affective design corresponding to products. According to the authors, pleasure and emotion with products are seen from three theoretical perspectives: (1) the context of use and activity, (2) categories of emotions with products including visceral, behavioral, and reflective, and (3) the centrality of the structure of human needs in the conduct of cognitive and affective assessment systems.

6.5.3 Methods for Evaluating User Experience (UX)

According to Vermeeren et al. (2010), the methods to assess the user experience are related to those used to assess usability. There is a subtle barrier that separates these two categories of methods. However, Rebelo et al. (2021) rightly point out that sometimes a good experience does not correspond to good usability. This is very clear, for example, in products of fashion design. A user may wear an elegant high heeled shoe that is fashionable, which will bring her great pleasure and a good user experience, but which may be uncomfortable, a bad usability. We would add that even then, the uncomfortable feeling will certainly not provide the user with a good experience at its highest level. On the other hand, a comfortable running shoe that presents a good usability may not bring a good experience to the user for not representing her status and lifestyle. Thus, Rebelo et al. (2021) conclude that usability can be independent of user experience.

The measure of satisfaction, commonly used in several methods to assess usability, is also considered a measure of the user experience. User experience is measured using subjective metrics (measures). A **metric** is a way of measuring or evaluating a particular phenomenon or thing. Metrics measure something about people and their behavior or attitudes. Thus, usability measures are usually objective (except when measuring satisfaction) and UX measures are usually subjective.

INTERNET

A list of about 90 tools for evaluating the user experience (UX) is available on the website below:

All UX evaluation methods, All about UX.
https://www.allaboutux.org/all-methods

According to (Leventhal and Barnes, 2007), user experience ratings are based on two typical usability studies:

- **Formative evaluation**: the UX designer evaluates a product or design periodically while it is being created, identifies deficiencies, makes recommendations, and then repeats the process until, ideally, the product comes out as close to perfection as possible. The authors make an analogy with a chef who periodically checks a dish while it is being prepared and make adjustments to positively impact the final result. It aims to make design improvements before launch. This means identifying or diagnosing problems, making and implementing recommendations, and then evaluating again.
- **Summative evaluation**: the UX designer evaluates a completed product in relation to specific objectives. The analogy regarding summative evaluation would be the chef evaluating the dish after it comes out of the oven. The objective is to define how much a product or part meets its previously defined objectives. Evaluation can also be conducted to compare multiple products against each other.

ATTENTION

Several tools of a subjective nature are used to evaluate the User Experience (UX) when handling products. Here are some based on Merizi et al. (2018) and Sacharin et al. (2012).

- **AttrakDiff** – This tool uses 28 pairs of opposite adjectives (semantic differentials) to evaluate the user's experience with a product. The semantic differentials are grouped into four categories: pragmatic quality, attractiveness, hedonic quality of identification, and hedonic quality of stimulation. Users evaluate the semantic differentials of the same or competing products using a seven-point scale (-3 to +3) (Hassenzahl et al., 2003).
- **Scenarios** – It is a tool that describes events of use of a given product, imagined by one or more users, through narratives (Rosson and Carroll, 2001).
- **Change Oriented analysis of the Relationship between Product and USer** (CORPUS) – It is based on retrospective interviews, with the objective of analyzing the changes in the nature of the user–product relationship, over a period of one to two years. The tool has five dimensions of quality, namely: utility, usability, stimulation, beauty, and identity of communication and global assessment. The user fills in a questionnaire about his perceptions about the product, according to each of the dimensions presented (Wilamowitz-Moellendorff et al., 2006; Hassenzahl and Tractinsky, 2006).
- **Day Reconstruction Method (DRM)** – It aims to obtain specific and recent memories, thus reducing errors and prejudices of remembrance. Participants fill out a "previous day's diary" where they describe each episode by answering questions about the situation (when, what, where, with, whom) and about the feelings they experienced, using a table with 12 affection descriptors (Kahneman et al., 2004).
- **Experience Sampling Method** (ESM) – It aims to record self-reports of a representative sample of moments in the users' lives. Users complete a questionnaire during a specific period during the day, for a period of time, a week, for example, during which reports of the user's experience with the product will be collected (Csikszentmihalyi and Larson, 2014).
- **Facial express recognition** – They are used to assess the individual's emotions by means of their facial expressions compared to previously established patterns for emotions, such as those described in the FACS – Facial Action Coding System (Ekman and Rosenberg, 2020).
- **Geneva Emotion Wheel (GEW)** – It can be used to evaluate products, events, and situations. The GEW is composed of a circle that is divided in half, horizontally, in two dimensions (positive and negative valence) and control (low to high). This forms four quadrants that are each composed of four emotions, totaling 16 emotions to be evaluated (Swiss Center for Affective Sciences, 2020).
- **Product Emotion Measurement Tool (PrEmo)** – It aims to study the relationship between the appearance of a product and the emotional reactions that it arouses in the user. The instrument brings 14 emotions, seven of them positive (joy, hope, pride, pleasant surprise, satisfaction, fascination and desire) and seven negatives (disappointment, fear, shame, unpleasant surprise, dissatisfaction, boredom, and disgust). These emotions are represented by images, so that the user can analyze how much the product's image aroused each one of those emotions (Desmet, 2002).

- **UX Curve and IScale** – They are similar methods. Users report, through a graph, how and why their experience with a product changes over time and how they relate to their customer loyalty from a time versus emotion perspective (Kujala et al., 2011; Karapanos et al., 2009).

As we have seen, the user experience is a reflection of the emotion felt in the interaction with the product. This opens the door for objective analysis to assess emotion as a method of analyzing usability and user experience, rather than traditional subjective methods. The use of **biofeedback** to evaluate the usability of products and, consequently, the user experience will be analyzed below.

NOTE

Want to learn more about usability and user experience? Read the following books:

Soares, Rebelo and Ahram (2021). Handbook of Usability and User Experience (UX): Methods and Techniques v.1, CRC Press. ISBN: 9780367357702.
Soares, Rebelo and Ahram (2021). Handbook of Usability and User Experience (UX): Research and Case v.2, CRC Press. ISBN: 9780367357719.
Albert, W.; Tullis, T. (2013). *Measuring the User Experience: Collecting, Analyzing, and Presenting Usability Metrics (Interactive Technologies).* Morgan Kaufmann. ISBN-13: 978-0124157811.
Rubin, J.; Chisnell, D.; Spool, J. (2008). *Handbook of Usability Testing: How to Plan, Design, and Conduct Effective Tests.* 2a. ed. Wiley. ISBN-13: 978-0470185483.

6.6   THE USE OF EMERGING TECHNOLOGIES TO ASSESS USABILITY AND USER EXPERIENCE

Usability assessments based on users' physiological responses (biofeedback) are beginning to be used by several researchers. Rebelo et al. (2021) advocate that virtual reality, integrated with biosensors, offers a strong way to assess UX through emotional reactions. Papers with tools based on neuroscience and infrared thermography are available in the specialized literature (Zeng et al. 2020; Soares et al. 2019; Barros et al., 2015, 2016; Ahram et al., 2016).

Next, we will analyze three main biofeedback tools: (1) Electroencephalography, (2) Eye Tracking, and (3) Digital Infrared Thermography.

6.6.1   Electroencephalography (EEG)

Neuroscience is a scientific discipline that is widespread beyond the borders of the medical field. Neuroscience principles and practices are applied in several areas, such as fine arts, architecture, design, engineering, music, computer science, linguistics, etc.

NOTE

Definitions:

- **Neuroscience** is the study of the nervous system, which includes the brain, spinal cord, and peripheral nerves.
- **Neuroergonomics** is the study of the human brain and its behavior at work. It focuses on the investigation of perceptual neural bases and cognitive functions, such as seeing, hearing, observing, remembering, deciding, and planning various technologies and configurations in the real world (Parasuraman and Rizzo, 2008).
- **Neurodesign** is a discipline based on design, psychology, neuroscience, and computer science, which aims to understand the user's experience from brain activities, investigating decision making from their interaction with the product or system.

Neuroergonomics and neurodesign work by collecting and analyzing information in the volunteers' brains. The tasks performed and emotions felt by users produce electromagnetic stimuli in certain regions of the brain that can be captured through electrodes located in caps and other devices. Electroencephalography (EEG) is an electrophysiological monitoring that is used to record electrical activity in the brain. Figure 6.6 shows three EEG models from the company Emotiv. Emovit Epoc + (a) has 14 EEG channels, is sensitive to the whole brain, and the battery lasts up to 12 hours. The Emotiv Epoc X (b) has similar characteristics to the previous one with a battery life of up to 9 hours. Emotiv Epoc Flex © is a 32-channel EEG cap with sensors that can be placed anywhere on the head or ears with included ear clips.

The tasks and emotions performed by users produce electrical activities in certain regions of the brain, which can be captured using electrodes. The electrode placement by the scalp is organized according to the international system 10-20 (Figure 6.7), which is taken as a standard for the acquisition of EEG signals and presents the correct electrode arrangement in the scalp, providing a high-quality EEG (Malmivuo and Plonsey, 1995).

This method is able to capture and identify discrete emotions, such as feelings of reward (pleasure, satisfaction), punishment (disgust, aversion), joy, fear, anger, and sadness. After the electroencephalographic signals are captured by the electrodes, they are analyzed and associated with activities and emotions using specific software, such as sLoreta (Figure 6.8).

Neuroscience allows the brain to control machines through EEG devices. This is the future of many consumer products and of immense assistance as assistive

FIGURE 6.6 Devices for acquiring EEG signals: Emotiv Epoc + (a), Emotiv X (b), and Emotiv Epoc Flex (c). Images kindly provided by Emotiv.

technology for disabled users. Among the various experiments already carried out with this technology, the one conducted by Miguel Nicolelis, a Brazilian neuroscientist, stands out. Nicolelis led a team of scientists to create an exoskeleton, controlled by the brain. This exoskeleton allowed a disabled quadriplegic to take the first kick at the official opening game of the World Cup, held in Brazil, in 2014. This feat was considered a milestone in the use of brain-controlled equipment. The great innovation is that the exoskeleton allowed the ambulation of quadriplegic patients and opened the door for projects of consumer products and assistive technology to be used in the future.

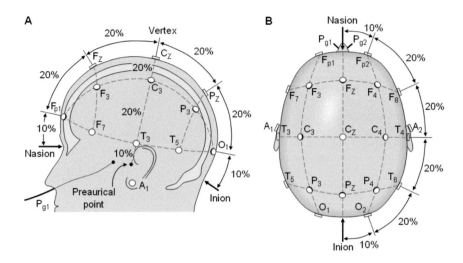

FIGURE 6.7 Electrode arrangement in the International 10-20 System (Malmivuo and Plonsey, 1995), with the permission of Prof. Jaako Malmivuo.

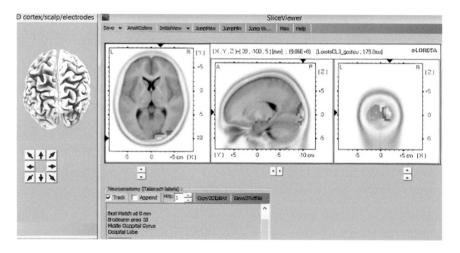

FIGURE 6.8 Image capture of brain activities through sLoreta software (Barros, 2016). Source: The author.

INTERNET

Do you want to know more about how EEG equipment works and how it can contribute to the brain-machine interface? What is the brain-brain interface? Watch these TED talks:

How does the brain work in everyday situations? Tan Le, 10 December 2014.

https://www.youtube.com/watch?v=HqtpMDEwiyM

Brain-to-brain communication has arrived. How we did it? Miguel Nicolelis, 2014.

https://www.ted.com/talks/miguel_nicolelis_brain_to_brain_communicatio n_has_arrived_how_we_did_it/discussion

6.6.2 Eye Tracking

The term "**eye tracking**" consists of a technique for understanding attention. The orientation of eye movements in relation to the individual's head or the position of the point fixed on a computer monitor, or another object, at a certain distance from the individual is an indicator of where the gaze should be directed. This is useful, for example, when designing information devices or to define the best position to distribute displays on a car dashboard. Eye tracking can be used through a device on the computer or installed on helmets or glasses (Figure 6.9).

Eye tracking reveals what the user is looking at when interacting with a product or system, helps to understand the area of attention and can be useful to assess the user's mental model. The method of capturing the focus or visual attention of the user occurs through a computer screen. Therefore, through eye fixations at the sampling sites, the Areas of Interest (AoI) are defined on the computer screen or scene seen (Goldberg and Wichansky, 2003). The relationship between eye movements and cognitive processing has been applied as a tool to improve the design and ergonomics of interfaces (Kramer and McCarley, 2003). Figure 6.10 shows the locations of ocular fixation in an analysis on the computer (a), in an analysis of packaging layout in a supermarket (b) and in a product analysis in a virtual environment (c). The use of virtual reality combined with eye tracking allows the designer to better understand the user's interaction with the product.

The areas of interest are used to identify user fixations that occurred in a particular area of interest for the experiment. The duration of the gaze points (gauze) are consecutive fixations in the same areas of interest and their duration is obtained by

(a) (b)

FIGURE 6.9 Examples of eye tracking equipment for tables and glasses. Images kindly provided by Tobii Group.

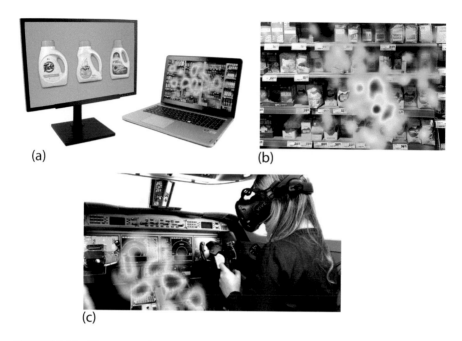

(a)   (b)

(c)

FIGURE 6.10 Heat maps of the eye tracking look in a notebook computer (a), supermarket (b), and virtual environment (c). Images kindly provided by Tobii Group.

adding the fixations (Jacob and Karn, 2003; Poole and Ball, 2005). The sum of the fixings produces **heat maps**. More information on eye tracking can be found in Eye Tracking (2020).

6.6.3   DIGITAL INFRARED THERMOGRAPHY

Digital infrared thermography is the use of an infrared thermal imaging system to detect, display, and record thermal patterns and temperature values on a given surface. All objects above absolute zero, or zero Kelvin (– 273,15 °C), emit infrared radiation that is detectable with a thermal imager, also known as an infrared camera. It is a non-invasive method that allows the capture of heat images not visible to the human eye, through a thermographic camera (Figure 6.11). The images, called thermograms, allow the researcher to analyze and identify the temperature variations of parts of the skin. The interpretation of the images requires the researcher to have the necessary qualifications in health science.

Figure 6.12 shows thermograms of an experiment conducted by Vitorino (2017), in which it is possible to verify changes in the temperature of the volunteers' arms and shoulders, before and after the activity of manipulating software through gestures, demonstrating fatigue. In these regions of the body, to conduct an experiment, the environment in which the test is performed must have a controlled temperature and the experiment protocol must be previously defined.

FIGURE 6.11 – Flir T1010 digital thermal camera. Images kindly provided by FLIR Systems, Inc.

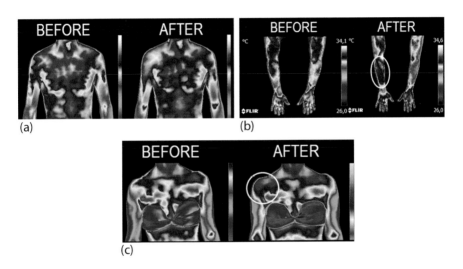

FIGURE 6.12 Thermograms of a volunteer before and after using a system in a gesture interface experiment. Thermograms showing temperature changes in: (a) the forearm of the volunteer, (b) the shoulders and arms of a male volunteer, and (c) the shoulders and arms of a female volunteer. (Vitorino, 2017). Source: The author.

Studies prove that thermography can be a good instrument to provide quantitative and physiological indicators to analyze emotions and can be used to replace subjective variables in usability studies (Vitorino, 2017; Barros, 2016; Marçal et al., 2016). Ioannou et al. (2014) and Merla and Romani (2007) carried out studies that prove the effectiveness of thermography in studies that evaluated conditions of emotional excitations such as empathy, stress, fear, sexual arousal, anxiety and pain and joy, through the capture of images of the face. According to the authors, the individual's response to an emotional stimulus may increase or decrease in the face, particularly in the nose, from certain emotional excitations.

Infrared thermography can be used as a tool to capture and analyze physiological changes applied to objective investigations on user–product interaction (Jenkins et al., 2009). The authors carried out a comparative study between thermography techniques, EEG and skin thermal conduction during the simulation of a cognitive activity. The study proved that it is possible to analyze the effects of affective experiences with thermography and correlate them with other measures obtained through electroencephalography, such as emotion, for instance.

The use of this tool in the analysis of human interaction with products is still scarce, especially when related to human psychophysiological parameters. Jenkins et al. (2007) conducted a study in which it was possible to measure the user experience during product interaction. The results of this study demonstrated that there were significant changes in the temperature of the participants' faces during the cognitive activity of assembling a puzzle.

The studies mentioned have shown that the thermography technique has a high potential to measure changes in the affective state during cognitive activities performed by product users. Thus, usability evaluations of the user–product interaction can use infrared thermography to analyze the user's affective state, becoming an objective technique in the usability evaluation, instead of the traditional subjective techniques.

Some experiments with Neuroergonomics, Neurodesign, Eye Tracking, and Infrared Thermography to assess the usability of products and software have been carried out under the supervision of the author.

Barros (2016) argued that, in usability assessments, the researcher does not always have the knowledge, in fact, whether the experience **reported** by the user at the time of the research fully matches his real experience **felt** during the experiment.

To verify this statement, the author carried out a pilot study in which the user interface handling PET bottle packaging is analyzed through:

- electroencephalography, in conjunction with eye tracking, to analyze the electrical activities in the volunteers' brains during the time of presenting PET bottles, associated with the command to fix the gaze on the bottle that the volunteer liked or disliked; and
- digital infrared thermography with the objective of capturing images of radiated heat from the volunteers' body parts to associate them with eventual emotional and/or actional excitations when handling the product.

As a result, the author concluded that in the same way that it is possible to analyze the **reported experience** by users through usability tests, it is possible to analyze the **felt experience** through the techniques of eye tracking, EEG, and digital infrared thermography. It has also been proven that these tools are effective in measuring the satisfaction (**felt experience**) of users while handling consumer products. This study was a pioneer in the use of these two biofeedback techniques (EEG and digital thermography), used together in the analysis of the usability of consumer products, and it is also very promising for ergonomic analysis of work situations.

Another study carried out by Vitorino (2017), under my supervision, evaluated the interface of a 3D modeling software that uses the movements of users' upper limbs (arms, hands, and fingers) through the gestural interface. The author analyzed ergonomic and usability issues using the Sculpting software controlled by the Leap Motion gestural motion sensor. Physical and emotional issues related to 12 experiment volunteers were investigated using digital infrared thermography for ergonomic and usability evaluation. The author was able to identify physical problems in the use of the gestural interface, recommending further studies before the industries adopted this technology as a standard in consumer products.

Rebelo et al (2021) present an integrated model to evaluate the User Experience based on their emotional reactions and behavioral decisions, using virtual reality and biosensor technologies. The authors analyze the use of biosensors to capture various physiological signals such as breathing, sweating, heartbeat, galvanic skin response, and brain activity. The authors present a low-cost electronic device in which multiple biosensors can be connected, and used to obtain physiological signals in the evaluation of the user experience.

REFLECT

In the previous chapters, you learned about the principles and practices of ergonomics and human-centered design. In this chapter, you learned about usability and user experience. Do you agree that the principles of user-centered design, usability, and UX are the same as those advocated by ergonomics?

6.7 DESIGN, ERGONOMICS, AND PANDEMIC

The emerging of the COVID-19 pandemic caused by the SARS-CoV-2 virus or new coronavirus radically changed relationships in social, leisure, and work environments. The world changed, and the coronavirus became an accelerator of the future, anticipating the discussion and implementation of issues such as remote work, distance education, the search for sustainability, and corporate social responsibility (Melo, 2020).

Facing this new world, the contributions of design and ergonomics must be considered within two contexts: the macroenvironment and the microenvironment.

MACROENVIRONMENT

The macroenvironment involves the individual's social and work relationships. Social isolation has imposed on individuals a routine of isolation and quarantine that has profoundly impacted social relationships. Distance work (including home office) and distance learning have become a constant in the lives of many workers, students, and teachers. The deprivation of social interaction in leisure, work, and study environments prevents professional and personal bonds from being continuously nurtured. It can be said that the levels of stress caused by the impacts of the pandemic and all its consequences, both in fatality numbers and in geographical scope, are unparalleled in the entire history of humanity and sometimes affect people's mental health.

The emotional imbalance caused by isolation, social restriction, and remote activities affects individuals' mental and physical health psychological well-being. Therefore, considering these elements associated with health is very important when referring to product design. No matter how good the design is, there will hardly be a good interaction of the Human with the Machine if there is no emotional balance during the execution of the work activities.

The World Health Organization defines mental health as "a state of well-being in which an individual realizes his or her own abilities, can cope with the normal stresses of life, can work productively and is able to make a contribution to his or her community" (WHO, 2021a). Mental health includes our emotional, psychological, and social well-being and affects how we think, feel, and act. It also helps us deal with stress, how we relate to others, and how we make our choices (Mentalhealth.gov, 2012).

The World Health Organization (WHO, 2021b) presents some recommendations that contribute to maintaining mental health and a social and working environment in times of pandemics. Among which we highlight:

- Prevent social stigma against a person, group of people, or people who share certain characteristics of a specific disease. This could lead to people being labeled, stereotyped, discriminated against, treated separately, and/or suffering the loss of status because of a perceived connection to a disease.
- Avoid watching, reading, or listening to news that makes you feel anxious or distressed.
- Take care of your basic needs by ensuring strategies such as rest during work or between shifts, sufficient and healthy food, physical activity, and maintaining contact with family and friends.
- Maintain social activity with relatives, friends, and co-workers through available digital devices.
- Use inclusive ways to share messages with people with intellectual, cognitive, and psychosocial disabilities.
- In case of isolation, try to maintain your routine as much as possible, including on social media.

MICROENVIRONMENT

The microenvironment involves the individual's personal relationships and emotional interactions with products. Therefore, overuse of digital products, technologies, and platforms is termed "Digital Dependence/Technology Dependence" (Arora et al. 2021; Rahayu et al., 2020; Singh and Singh, 2019; UNBC, 2021). Digital dependence occurs when it produces impairment in social and family relationships and can also bring physical consequences, including back pain from long periods of sitting, tendinitis, repetitive strain injury, and mental problems such as depression and anxiety.

Factors associated with deteriorating physical and mental health are identified in digital addiction, such as eye discomfort, posture problems, sleep problems, feelings of guilt, anxiety, depression, isolation, and agitation (ICANotes, 2018; Johnson, 2020). In addition, digital addiction can generate "nomophobia" which is the irrational fear of being without a cell phone or electronic devices in general. Although digital addiction has not been formally recognized as a mental illness, some research reveals that it has symptoms similar to those identified with psychoactive substance dependence (Yen et al. 2008, Brand et al. 2014).

In pandemic times the use of digital communication devices (computers, smartphones, and tablets) was intensified (Arora et al., 2021). According to the Pan American Health Organization (PAHO, 2021), increased global access to Internet-connected cell phones, as well as social media, has led to the exponential generation of information creating an information epidemic, or infodemic (Arora et al. 2021). For example, the authors reported that in China and the Philippines smartphone use in the COVID-19 pandemic period grew by 86% and internet use worldwide grew by 70%. The use of digital technologies in the pandemic period increased the risk of digital addiction by two or more times, and constant use of the internet and such devices associated with social restriction, work, and remote study significantly increase mental illness (Arora, 2021; Tripathi, 2018).

Mental problems associated with excessive use of the internet and other digital technologies, including smartphones during the COVID-19 period, have been studied by several authors (Király, O. et al, 2020; Gao J et al., 2020; King et al., 2020; Sun, Lin and Operario, 2020). Potas et al. (2021) studied the impact of technology dependence on 382 adolescents in the COVID-19 era and confirmed the consequences of the effects associated with physiological and psychological repercussions. How the pandemic may affect the potential for skill and habit development in the generation of youth, children, and adults is a question yet to be investigated.

The ergonomic design of digital devices can minimize some effects related to the physical and mental health of individuals ensuring, among others, better usability and reduced mental load in human × machine interaction. However, the lack of proper ergonomic design in digital devices can lead to musculoskeletal disorders such as hand, wrist, arm, neck, and back pain, caused mainly by repetitive motions and inappropriate postures. The Chartered Institute of Ergonomics and Human Factors (2021) has developed a guide that provides recommendations for sustainable work using digital equipment. This guide is available for free download.

The Japan Human Factors and Ergonomics Society (2020) has also prepared a guide called "Seven Practical Tips for Teleworking/Home-Learning using Tablet/Smartphone Devices". The recommendations are:

- When using digital devices for telecommuting or learning tasks at home, every 20 minutes of activity, pause for 20 seconds, looking at an object that is 20 feet away.
- Alternate sitting and standing postures when using a tablet and/or laptop computer.
- When using a smartphone, keep the device about 40 cm away from your eyes and, in order to avoid fatigue, support the elbow of the arm on which you are using the smartphone with the opposite hand.

- Keep the tablet, smartphone, or notebook at or slightly below eye level. The use of assistive devices or even in their absence a stack of books is recommended.
- When reading on a smartphone or tablet screen, always choose the "landscape" option in order to improve usability and reduce eye fatigue.
- Take micro-breaks when using tablets or smartphones.
- Use an external ergonomic keyboard when typing for a prolonged period.
- If you have an air conditioner or fan, make sure the airflow is from top to bottom and back to front.
- Avoid excessively bright lamps, reflections on the mobile device screen, or working in the dark.
- Whenever possible, keep the temperature around 21°.
- Adjust the sound of your digital equipment and environment to a comfortable level. Researchers recommend 80 dB. If necessary, you can download an application to measure environmental noise (e.g., Sound meter, Decibel X, Decibelimeter Pro, and dB Meter). Some of these applications are free.
- Create an emotionally stable work and/or study environment. Talk with your family about the space and time needed for your study or work. Don't forget to set aside some time for your family as well.

Positive emotions are essential for a good user experience and as such one of the functions of design (this topic will be further explored in the next chapter). With the outbreak of the COVID-19 pandemic, people started to wear masks with props for protection and prevent the virus's spread. People began creating masks with logos of famous brands like Chanel, and Gucci encrusted with jewelry as a way to maintain their self-esteem (FastCompany, 2021).

Regarding the disabled user, the Research Institute for Disabled Customers (2020) conducted several types of research involving inclusive design and accessibility for the disabled useful for the pandemic time, such as impacts and restrictions of COVID-19 on disabled users, accessibility and infrastructure for electric cars, car trunks to accommodate wheelchairs, cell phone apps, etc.

Issues related to the role of the designer in the post-COVID world are beginning to be discussed. From Betsky (2021) and Gensler (2021), we present some questions that should be answered sooner or later with the contribution of designers and ergonomists. They are: how to develop products using sustainable materials and safer ways of use?, how to adapt gesture controls to various digital equipment used in everyday life?, how to design public spaces that can restore people's confidence?, how to make workplaces more accessible?, how to reorganize work according to the "new normal"?, how to design new forms of communication in the future?, how to develop new ways of digital health for non-face-to-face care and remote medical procedures?, how to create new ways of distance learning? These are just some of the questions that will have to be answered in the near future.

In conclusion, we state that product design and ergonomics can contribute to improving the physical and mental health of individuals in this pandemic and post-pandemic period. This can be done by providing better usability of products and digital platforms, developing sustainable products, improving workspaces, and building environments, in general, to provoke positive emotions in users. To do this, sustainable and natural materials (such as woods and natural fibers) should be incorporated into the design of products and environments and contribute to work organization and tasks that consider users' physical and cognitive limits and abilities, etc.

NOTE

Do you want to know more about remote work and specific recommendations about products used in the home office? Check it out: 26 WFH Tips While Self-Isolating During the COVID-19 Outbreak (Healthline, 2021); How to set up a WFH 'office' for the long term (Gruman, 2021); Telework resources during the COVID-19 pandemic (Office of Financial Management, 2021); Teleworking during the COVID-19 pandemic and beyond: a practical guide. (ILO, 2020).

The next chapter presents the **Ergodesign Methodology for Product Design** based on the knowledge studied so far: analysis of the product design process, how to meet the needs and satisfaction of users, how to design for disabled users, and how to incorporate methods based on users' needs for the design and manufacture of consumer products.

6.8 SUMMARY OF THE CHAPTER

- The rapid growth of a very competitive market requires quality in all aspects of a company's operations, with things being done right first time and defects and waste eradicated, as much as possible, from operations.
- User satisfaction is the company's highest priority and is obtained by providing a high-quality product and continuously improving the quality of the product to maintain a high level of customer satisfaction.
- User needs and product specifications are useful for guiding the conceptual phase of product design.
- Quality function deployment is a method that translates the customers' demands into design targets and major assurance points to be used throughout production.
- The House of Quality, the first of a series of translation matrices used in the QFD method and the most important from the point of view of ergonomics and product design, links customer need to the desired and specific product characteristics.
- The process of QFD continues after finishing the House of Quality. Other linked houses should be used to convey the user's voice through to manufacturing.
- We have examples of poor usability in many products that we find in our daily lives. There is no product with good design and poor usability.

- ISO defines usability as "the extent to which a product can be used by certain users to achieve specific goals with effectiveness, efficiency, and satisfaction in a certain context of use". Jordan examines the concepts of effectiveness, efficiency, and user satisfaction.
- Usability requires that the product or system has an efficient use, is easy to learn and remember, does not cause errors, promotes pleasure in use, and guarantees user satisfaction.
- There are several usability models, most of which are proposed for software environments, although they can also be applied to product design. Jordan (1998a) presents a specific usability model for products, and Leventhal and Barnes (2007) present a hybrid model based on classic authors.
- There are many tools used for evaluation and usability tests. The usability research data can be divided between quantitative and qualitative data.
- Usability assessments based on users' physiological responses (particularly obtained through Neuroergonomics, Neurodesign, and Digital infrared thermography) are beginning to be used.
- The user experience is the result of the user's interaction with the product or system (interface) provided by usability.
- User experience is measured through usability metrics (measures) and is based on formative and summative assessments.
- Emotional design is the characteristic that a product has to arouse emotions in end users, through its design, in order to establish connections between the product and the user.
- Norman relates the emotional process to three levels of human brain processing: visceral, behavioral, and reflective.
- Emerging technologies, based on biofeedback provided by EEG, eye tracking, and infrared thermography, present themselves as an alternative for evaluations of usability and user experience, complementing traditional subjective techniques.
- What is called human-centered or user-centered design, usability, and user experience have always been part of the principles and approaches of Ergonomics.
- The contributions of design and ergonomics in the pandemic and post-pandemic periods must be considered within two contexts: the macroenvironment (which involves the individual's social and work relationships) and the (microenvironment which involves the individual's personal relationships).
- In the pandemic and post-pandemic period, product design and ergonomics can contribute to improving the physical and mental health of individuals by proposing products and digital platforms with better usability, developing sustainable products, better workspaces, and environments in general.

7 ErgoDesign Methodology
A Human-centered Methodology for Product Design

This chapter is dedicated to the presentation, analysis, and detailing of the Ergodesign Methodology for Product Design. Initially, some general considerations are presented that are useful for understanding how the methodology was formulated and its major objective, which is to incorporate the "voice of the user" as method of co-design in the design process. An electric wheelchair, for internal and external use, is used as an example in all the proposed methodology. The stages of the methodology in which the designers are involved are explained in detail: Approaching the User, Investigating the Problem, Product Planning, Design Creation, Prototyping, and Testing and Verification. The innovative User Panel is presented in detail. The use of Quality Function Deployment (QFD) as a tool to meet users' needs and product functions is described in detail. Several techniques for generating new ideas are described, as well as formal techniques for filtering and selecting such ideas. Recommendations for user manual design and product prototype evaluation, verification and testing are presented. The validation of the methodology proposed by a sample of designers is outlined, resulting in a revised version of the proposed methodology. Finally, a step-by-step summary and graphical representation of the methodology steps are presented.

7.1 GENERAL CONSIDERATION

The previous chapters have revealed some important findings that indicate the need for a human-centered methodology for product design. Also, extensive surveys of designers were carried out, particularly designers specializing in designing wheelchairs, prescribers (physiotherapists and occupational therapists), rehabilitation engineers, users, and carers on their views on wheelchair design, assessment, prescription, and wheelchair use. These surveys were carried out during the author Ph.D. course at Loughborough University, United Kingdom (see Appendices 2 to 6).

It was found that most designers in the survey carried out all phases of the design process based on their assumptions about users' expectations and needs and did not listen to "the voices of users, carers or prescribers" in the design process. It seems that this reality is not much different for the designers of consumer products, even today.

Thus, it is important to bring together the major features of the results of the field studies carried out with the stakeholders involved in the processes of wheelchair

DOI: 10.1201/9781003214793-7

design, supply, prescription, and use, in an attempt to highlight the deficiencies to be overcome by the human-centered methodology. The features described below are only those who may provide some kind of contribution to the production of the methodology for wheelchair design on a mass-production scale. It is important to call attention, once more, to the fact that although this methodology was originally designed for wheelchair design, it can be used for any consumer product.

We draw attention to the fact that, even though it was carried out many years ago, the research carried out by the author is justified as a report of the importance of hearing the "voice of the direct users", and in this case, also the "indirect" ones. This importance is demonstrated by the wealth of contributions that can be translated into requirements for product design. Using the "user's voice" in design requirements is the foundation of the proposed methodology. Furthermore, we believe that the lessons learned from this research are useful for those interested in product design for the disabled. The summary of the report of the research participants can be found in the Appendices 2–5.

The methodology discussed below aims to overcome the discrepancies highlighted in the research carried out by the author with users and incorporate the "voice of direct and indirect users" in the design process. This approach is called **co-design**. It is important to draw attention to the fact that some aspects, such as costs and manufacturing processes, are not presented in depth. The methodology will focus on aspects related to the design, ergonomics, and usability of the product in order to meet the needs of users. Again, we point out that, although this methodology was originally proposed to be used in wheelchair design, it can be adapted and used for any consumer product in general.

7.2   THE ERGODESIGN METHODOLOGY FOR PRODUCT DESIGN

The proposed methodology is intended to be used for the design of any consumer product. This means that the product must cover the widest possible range of users, and the methodology must be adapted to the particularities of the product for which it is intended. This methodology can also be adapted for use at the university as a way of teaching students how to use design methodology.

NOTE

It is important to clarify some definitions (based on Marconi and Lakatos, 2017).:

- **Methodology** is a set of methods, techniques, and tools.
- **Methods**, in turn, are logical procedures to achieve a specific objective. These are sufficiently general procedures to enable the development of scientific research.
- **Techniques** are sets of precepts or processes used by a science or art, in addition to being the ability to use precepts or standards in practice.
- **Tools** are resources used in the application of the methods. Usually, a tool is associated with a concept and/or method.

It is also the intention of this methodology to provide a step-by-step guide to be used by the designer in a way to assist him/her in: (a) making decisions about the several design dilemmas throughout the distinct design phases, reducing the possibility of moving forward with unsupported decisions and allowing the other members of the team to understand the decision rationale; (b) obtaining information on the use of a variety of data gathering techniques; (c) taking a series of key steps assuring that relevant design issues have been considered in the design process and (d) organizing documentation of the various design phases in order to facilitate the decision-making process, to be used for future reference and for educating new members of the design team.

In this methodology, the design team includes industrial designers and ergonomists. Other professionals such as mechanical, manufacturing, and production engineers, finance, marketing, and sales personnel, and the management team should interact with the design team in the several phases of product development.

The proposed methodology must be seen as a dynamic entity, capable of being modified and accepting continuous improvement. This means that some of its stages can be adapted and modified to meet the characteristics of the product for which it is intended and the organizational characteristics of the company in which it will be used.

It will take a powered wheelchair for indoors and outdoors use as an example to be used throughout the methodology. The example and the project situation are fictional with the unique purpose of illustrating the various steps of the methodology and situations faced by the designers. Although a number of appropriate techniques to assist the designers in different phases of the design cycle are shown, a range of others may be applicable and can be alternatively chosen for a specific situation. This will be mentioned when appropriate. Also, it is important to draw attention to the fact that although the methodology is intended to be used for wheelchair design, its use for other consumer products will depend, of course, on being adapted to the intended situation.

Wheelchairs can be understood in the context of the current methodology as a **system**. According to Chapanis (1996), "a **system** is an interacting combination, at any level of complexity, of people, materials, tools, machines, software, facilities, and procedures designed to work together for some common purpose". Thus, wheelchairs are systems that include the product itself, wheelchair users, and their carers. The "system wheelchair" is divided into "subsystems", for instance, "subsystem of seating", "subsystem backing", "subsystem of movement", "subsystem of braking". In its turn, the "subsystems" may be made up of still smaller units here called "components". So, in the "subsystem of movement", for example, there are a number of identifiable components such as the wheels, the frame, the engine, the connections. Some of the subsystems and their components are part of the industrial designer intervention. Others are related to the activities of mechanical or electrical engineers. A summary of the methodology is presented in Appendix 1.

The **Ergodesign Methodology for Product Design** comprises a set of 11 phases presented in the next sub-chapters. Figure 7.1 shows a flowchart illustrating the main phases of the methodology. In this figure the participation of the **User Panel**,

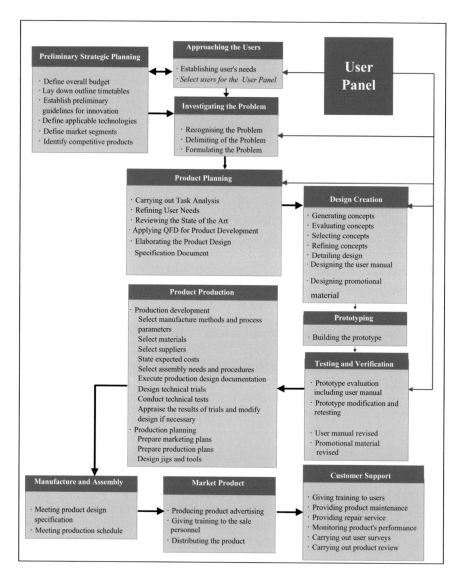

FIGURE 7.1 The first version of the Ergodesign Methodology for Product Design. Source: The Author.

described below, is with blue lines and arrows. A summary of the methodology is presented in Appendix 1.

The first stage of the methodology is the **Preliminary Strategic Planning**. Several actors are involved in this stage, including the designers, albeit indirectly. Designers are directly involved in the following five phases, including User Approach, Investigation of the Problem, Product Planning, Design Creation, Prototyping, and Testing and Verification. As they are part of the design activities, these phases are analyzed in detail in the methodology. The rest of the steps, referring to the Production, Manufacturing and Assembly Processes, Product Marketing, and Customer Support, will not be discussed in this book as they are not directly related to the activities of the designer or ergonomist.

7.2.1 PRELIMINARY STRATEGIC PLANNING

This first phase of the product development method usually involves a series of decisions taken by the company directors, which start the whole process. This early stage usually has a very limited direct involvement of the designers and will, therefore, only be briefly described in this book.

This phase comprises a number of strategic decisions, including:

- The definition of a **business plan** for the new product, indicating that it will, for instance, present a good business opportunity and sell in sufficient numbers to exceed its development costs and yield other economic benefits.
- The identification of the **relation between the new product and the company's other products**, indicating that the new one will, for instance, offer product users a clear benefit over existing products and that there will be significant product differentiation between the new one and its competitors.
- The definition of the **costs** associated with the expenses of the product development program.
- The establishment of a **timetable** for the product development process is defined by the time between the instant the first person starts to work on the product development program and the instant the final product is available to the user.
- The establishment of **preliminary guidelines for innovation**, as a result of a decision involving the company's business plan and the product's intended position in the marketplace in relation to competitors.
- The definition of applicable **technologies**, taking into account the current technology available in the company, and the need to identify relevant and emerging technologies responsible for product innovation.
- The identification of the **target market** for the product meaning the analysis of the business opportunity in terms of which market segment the product is aimed to capture and preliminary planning for its future commercialization inside and outside the country.
- The identification of **competitive products** representing the discovery of which products and their features with similar characteristics are currently in the market place. This will provide data for further analysis and evaluation of competitive products. This phase may be carried out by designers themselves and/or the marketing personnel. It includes the collection of promotional materials launched by competitors and maybe the acquisition of competitor products for future analysis and evaluation.
- The selection of a **sample of users** for the **User Panel** is a step carried out simultaneously with the next phase of the design process. It has the strong

participation of designers and, in view of this, will be explained and discussed in detail in the next sub-chapter.

7.2.2 APPROACHING USERS AND OTHER STAKEHOLDERS

One of the most important aspects of the product development cycle is to understand and learn from the user and the other stakeholders (indirect users, resellers, and suppliers). Understanding the needs of users is absolutely fundamental to identifying, specifying, and justifying a feasible design of the product. The users, in this case, comprise direct users (the product users themselves) and indirect users (in the case of wheelchairs, the carers).

This phase of approaching users will be divided into the following steps:

- Investigating **existing information** about product direct users and their indirect users;
- Developing **profiles** of product users and other stakeholders;
- Contacting **product direct and indirect users**;
- Selecting product direct and indirect users to **participate in the consulting process**;
- Carrying out **focus groups** with product direct and indirect users; and
- Selecting a **sample of users** to take part in the **User Panel**.

7.2.2.1 Investigating Existing Information about Product Users

As previously mentioned, this methodology is intended to be used in the development of a mass-produced product. Consequently, it can be assumed that it involves a company with reasonable experience and some time running the business in the marketplace. Thus, the first step in surveying product users should be the design team investigating existing information about the users inside the own company, such as:

- The company sales records, including repair and replacement parts;
- The register of complaints;
- Product warranty data; and
- List of former and current company's customer and their indirect users.

This will help designers (a) to have an overview of the product produced by his/her company, (b) develop a profile of the actual and potential users of the product, and (c) produce a preliminary list of product direct and indirect users who could be contacted.

7.2.2.2 Developing Profiles of Product Users and Other Stakeholders

In this stage, the following will be: (a) defined the profiles of the direct and indirect users who will participate in the focus group sessions, (b) contacted the users, (c) defined how many users will participate, and (d) recruited and selected the participants to compose the **Users Panel**.

(a) Defining Profiles of Direct and Indirect Users

An ideal scenario for the participation of direct and indirect users in the design of products should be to involve in the design process everyone who uses or is going to use the product. This will help to ensure that the product will attend to everyone's needs. Thus, it is necessary to choose a sample of participants whose profile is representative of those of the intended end-user population.

Ideally, the company will have developed a profile of the actual and potential users of the product long before the time for the development of the new product. However, if it has not been done yet, the design and/or marketing team will have to do it as one of the first steps in the development of the new product.

(b) Profiles of Product Direct Users

As far as consumer products are concerned, the end-user population comprises a large range of users with diverse shapes, dimensions, and needs. In developing a profile of users, the design and/or marketing team should capture a number of different characteristics, including:

- Age;
- Gender; and
- Education level.

If it is a product for a disabled user, should also include:

- Nature of disability (e.g., arthritic condition, amputation, respiratory condition, aging);
- Physical limitation (e.g., lower and/or upper limbs);
- Other(s) limitation(s) (e.g., visual, hearing, cognitive and/or verbal) and other(s) problem(s) (e.g., coordination of movement).

Additionally, if he or she is a wheelchair user should include:

- Quantity of wheelchair(s) each user has;
- Length of time of using wheelchairs;
- Type of wheelchair(s) the users own (e.g., manual self-propelled wheelchair, manual attendant propelled wheelchair, powered indoor/outdoor wheelchair); and
- Source of supply of user's wheelchair(s) (e.g., public or private market).

This information may also be part of a company database. It will be useful for defining a company's strategy in the launching of future products and an on-hand source for selecting and recruiting users to take part in user trials.

It is important to have in mind that for some particular data gathering tasks from users, e.g., focus groups and discussion groups, a sample of "typical users" is more appropriate to be used than a "representative sample" of the wider population of product users. **Persona** is a method complementary to user analysis, activity, and user context. It is a method that represents an abstraction of target users who share common behavioral characteristics (i.e., is a hypothetical archetype of real users). Brangier and Bornet (2011) and Miaskiewicz and Kozar (2011) introduce a good review of the Persona method.

(c) Profiles of Other Stakeholders

Apart from the direct users themselves, there are other stakeholders involved directly or indirectly with some kind of products. Wheelchairs, for instance, have other stakeholders such as carers (indirect users) and the service and product provider.

Establishing the profile of the indirect users is almost, if not equally, as relevant to a human-centered design as specifying the characteristics of the product users themselves. Important data to be included in indirect user profiles should be:

- Age;
- Gender;
- Education level;
- Time the carer spends assisting the product user; and
- Physical limitations.

(d) Contacting Direct and Indirect Users

In addition to contacting some product users included in the company's database, users of products manufactured by competitors must be recruited to guarantee the participation of users with different experiences, viewpoints, and knowledge in dealing with the product.

(e) Participation of Direct and Indirect Users

The participation of direct and indirect users in the **Ergodesign Methodology for Product Design** aims to: (a) produce inputs for the Preliminary Strategic Planning; (b) identify the needs of the direct and indirect users of the product; and (c) involve, in a participatory way, several direct and indirect users of the product in the design process.

It is important to keep in mind that consulting users to identify their needs is not a research task in academic terms. The goal of involving users in product development is to discover the most serious problems that users are likely to have when using the product and to get suggestions for incorporating their needs and wants into product design.

(f) Deciding How Many Participants to Include

Defining how many users should be approached in the identification of user needs is a matter that strongly involves monetary costs. The answer to this question has been addressed by a number of authors, including Nielsen Norman Group (2019a, b), Albert and Tullis (2013), Rubin and Chisnell (2008), Dumas and Redish (1999), Caplan (1990), Griffin and Hauser (1993), and Virzi (1992). Depending on the method used to approach the users, the authors generally agree, as a practical guideline, that conducting fewer than ten interviews is probably inadequate, and 50 interviews may be too many. Nielsen Norman Group (2019a, b) defends that testing with five people lets you find almost as many usability problems as you would find using many more test participants.

In terms of focus groups and discussion groups, with a number of fewer than six people, ideas and interactions may spar, and the group may be monopolized by one or more talkative people. With more than ten, the group may be more difficult to control and to guarantee adequate participation by each group member. Therefore, six to ten people may be ideal for conducting each focus group session. The number of participants is influenced by the availability of money to cover the costs, the time available to run this phase of the project, and the scope and depth of information to be obtained.

The use of focus groups has been revealed as being an appropriate tool in obtaining information about users' opinions, attitudes, preferences, and self-reports about their performance when using the product (Leahy, 2020; Langford and McDonagh, 2003; Bruseberg and McDonagh-Philp, 2002). In this methodology, a focus group is the method recommended to be used in the early phase of the design process and to establish initial user needs.

A number of three to five focus group sessions are suggested. Each session should include between six and ten people in total, with a ratio of about four product direct users to one indirect user if this is the case. This represents a total of about 15 to 40 product direct users and 3 to 10 indirect users to be recruited considering three or more sessions.

(g) Recruiting Participants

The company's own database is the primary source to find participants to take part in focus groups. As was previously mentioned, it is also essential to have users of competitor companies as members of the sessions. Although the participation of eventual product users employed by the company in the focus group session can apparently reduce the costs of the sessions, this is not recommended because: (a) they may feel intimidated and not criticize the products manufactured by their own company; (b) there may be difficulty in obtaining their release for the participation due to company's hierarchy and internal issues; and (c) the company may lose more money by releasing them to take part in focus group sessions than having them do their ordinary jobs.

The main mean of finding participants will be advertising in local newspapers, supermarkets, schools, universities, and/or Community Centers, special interest groups, social media, or hire a specialized company that can do this service.

Ideally, depending on the company's interest and availability of funds, the focus groups may be carried out in different regions of the country to improve the chances of obtaining a variety of responses and point-of-views from people of a different lifestyle, living environments, and different climatic conditions.

Apart from arranging transport to and from the meeting(s), the company has to budget for payment or other incentives for the participants, as gifts, for instance. If some disabled people take part in the focus group, the accessibility of the location must be considered in detail, including not only the room where the meetings will occur but also adequate access, the immediate environment such as toilets, ramps, lifts, the lunchroom, and the table that the participants sit around.

7.2.2.3   Selecting Product Direct and Indirect Users to Take Part in the Focus Group Sessions

There are no specific requirements for the participants, apart from having experience as the product under analysis and the ability to communicate verbally. The participation of "typical users", covering the extremes of the population such as fat man, thin woman, young user, elderly user, experienced user, novice user, etc., is recommended. He or she must have experience with the type of product being studied.

7.2.2.4   Carrying Out Focus Group Sessions with Product Direct and Indirect Users

It is recommended that the extent of each focus group session in this presented methodology should be 2–3 hours. The major components of focus groups are the facilities, the moderator, the participants, the procedures, and results. They are described as follows based on Cyr (2019), Krueger and Casey (2014), and Gatti (2012).

(a) The Facilities

The facilities can be understood in this context as the link between the company, the moderator, and the participants in the groups. Good facilities perform an essential role in the success of the focus group, which includes the following actions:

- Confirming participants' attendance and providing adequate transport.
- Providing a reception area for greeting participants.
- Guaranteeing a large and comfortable meeting room with easy external access for the participants and enough internal space to permit participants' movements with their products.
- Conducting the participants to the focus group room and accommodating them comfortably around a large table in a way that their products are placed under the table.
- Establishing a designated location for each participant to sit around the table with visible name cards on the table to promote interaction among participants and help moderator communication.
- Guaranteeing, if appropriate, extra space in the focus group room for the analysis of competitors' products, models, or prototypes.
- If possible, guaranteeing an observation room where the company's director(s) and members of the design and marketing team can see the proceedings easily. This is usually done using a one-way mirror room. Alternatively, a TV close circuit may be used. It is important to be sure that the observers are comfortable, and the sound system must allow each of the participants to be heard, even if they speak in a very low voice. It is important that volunteers know that they are being watched.
- Providing, if required, a display panel for figures, photographs, or other visual stimuli in the focus group room.
- Providing a computer with projector, whiteboard, or a flipchart and markers in the focus group room so the moderator can make notes if necessary.

- Providing appropriate support materials, e.g., photocopies, pencils, and notepads.
- Providing audio and/or video recorder support.
- Providing food and drinks for the participants and observers.
- Assuring that the materials used during the sessions are returned to the company in a way to keep confidentiality.
- Providing remuneration or some other type of compensation to the participants.

(b) The Moderator

A considerable part of the success of a focus group lies in the hands of the moderator. The session moderator should be familiar with the objectives to be achieved by the focus group sessions and should understand the product or problem being discussed.

Skilled moderators can ensure that:

- A good atmosphere is created.
- Rules for the discussion are established.
- The discussion is directed along relevant lines.
- Disruptive behavior on the part of specific participants is avoided or curtailed.
- Individuals, ideas, and ideologies are protected.
- All participants get an opportunity to contribute, and the proceedings are not dominated by any one person or group.
- Bias is eliminated so far as possible from the findings.
- The degree of probing and the depth of insight are sufficient to accomplish the research objectives.

Unskilled moderators find themselves conducting individual interviews with each of the participants rather than stimulating interaction within the group. If no member of the design or marketing team has experience in running focus groups, it is strongly advised to hire the service of experts in the field.

The role of the moderator in the focus group should be specified in terms of preparation, implementation, and post-group procedures as following:

(c) Preparation (to be carried out with company director(s) and the design and marketing teams)

- Developing research objectives.
- Defining the criteria for participant inclusion in the groups.
- Determining the number of focus group sessions needed to achieve the research objectives and the facilities for running the sessions such as access to the room, food requirements, instruction to the members.
- Deciding on the use of pictures, prototypes, or product samples to be analyzed; establishing concept statements and demonstrating ideas.
- Producing a guide for the moderator, including the set of questions to be addressed, the timing of various topics, and the use of external stimuli.

(d) Implementation

- Ensuring that the right people participate in the sessions.
- Briefing the company's personnel, who will observe the groups, on the objectives of the session and the content of the guide for the moderator.
- Conducting the group to cover all the elements in the guide provided to the moderator.
- Finishing the session(s) on time.

(e) Post-group Procedures

- Obtaining audio- and/or video recorder that was made with the groups.
- Analyzing the results.
- Producing a report explaining the findings and their consequences.

It is important to ensure that if the moderator is a person outside the company, he or she should be aware of any information that will enhance the effectiveness of the focus group, including strengths and weaknesses of the company and competitor's products and new ideas and concepts that may be explored.

It is advisable to consider the use of an "assistant moderator" to help the moderator with some tasks such as taking comprehensive notes, operating the audio- or video recorder, handling the environmental conditions and logistics (refreshments, lighting, seating, etc.), and responding to unexpected interruptions.

(f) The Participants

Additional attention should be given if there are any disabled participants in the sessions. It is important to guarantee their comfort and easy movement around the room if they use wheelchairs. Details of the participants' involvement were previously described when the role of the moderator was explained. Other recommendations concerning the involvement of the participants in the focus groups sessions include that they:

- Should speak as clearly as possible, one at a time, and facing the audience.
- Ask the moderator to clarify or repeat the question if necessary.
- If appropriate, feel free to make any comments and express eventual dissatisfaction regarding any of the products which are being discussed.
- Respect others' opinions even if they strongly disagree and follow the normal rules of polite conversation.

(g) The Procedures

The first moments in focus group discussion are critical and may be responsible for its success. Excessive formality and rigidity may inhibit and/or restrain interaction among participants. On the other hand, too much informality and humor can lead participants to not take the discussion seriously. The responsibility for creating a good atmosphere will depend basically on the moderator.

The sessions procedures should include:

- Introduction - The moderator introduces him or herself to the participants, briefly explaining the purpose of the session, alerting the participants that the session is being audio- and/or video-recorded, the existence (if it is the case) of a one-way mirror, the ground rules of the session and, finally, asking participants to introduce themselves.
- The core: The participants are asked to discuss the issues related to the topic, are guided to identify important information about the products under analysis, including their feelings and needs, the strength and weaknesses of the product, and to make suggestions about how to improve the design of the product. Special emphasis must be given to the establishment of user needs.
- Summary: The participants have the opportunity to share any information about the topic that they may have forgotten or otherwise omitted.
- Debrief: The moderator should finalize the session by thanking the participants and hand out the honoraria or other compensation.

(h) The Results

Analyzing the findings of focus groups is a very time-consuming activity comprising the transcription of hours of audio- and/or video-recorder. It involves a systematic analysis to gather and handle the data in a form useful for the design activity. The analysis must be verifiable in a way to permit another researcher to arrive at similar conclusions using the same documentation and raw data.

The researcher must have the skill to select and interpret the data focused on the study from the large amount produced in several sessions. The analysis process involves consideration of (a) the words used by participants and their meanings, (b) the context of the discussion including the tone and intensity of the oral comment, c) internal consistency as a result of changing positions, (d) extensiveness, frequency, and intensity of some comments, (e) specificity of responses based on experiences and (e) finding based on good ideas emerging from some evidences – words used, body language, the intensity of comments – rather than from isolated comments alone.

ATTENTION

The focus group findings should be stated in the form of a report and should include:

- A description of the purpose of the study;
- The number of focus group sessions held;
- A description of the focus group sessions;
- The methods of selecting and recruiting participants;
- The number of people in each focus group;

- The results, their interpretation, and consequential actions; and
- The appendix including, for instance, the questioning route for the focus group, the screening questionnaire, additional quotations, and a list of the product user needs to be classified by categories of products components (e.g., for a wheelchair: seat, back, handles, cushions, etc.).

It is important to draw attention to the need for data reduction for both the analysis and the quotations to be included in the report. The report should be of a reasonable length and not overlong. Thus, one should try to summarize as much as possible and avoid redundancies.

As shown in Figure 7.1, **Approaching the users** is an activity that interacts in a loop with **Preliminary Strategic Planning**. This means that the **Preliminary Strategic Planning** defines a series of queries to be discussed by the focus groups and is expected to receive a number of feedback, including the identification of strengths and weaknesses in some product models, the establishment of user needs, and so forth.

7.2.2.5 Selecting Members and Define Actions of the User Panel

The **User Panel** is a selected group of users who will participate, in a participatory way, in the various stages of the **Ergodesign Methodology for Product Design**. This group will have the function of acting, in a participative and consultative way, with the designer to assist in the various dilemmas and design solutions (co-design). Next, we will present ways of implementing this panel.

(a) Selecting Members for the User Panel

As previously commented, the use of focus groups is an excellent technique for evaluating concepts, identifying issues, and determining users' attitudes about products. However, the focus group approach is established on a consultative basis. For instance, the design and/or marketing team and the moderator choose the topics for discussion and the criteria against which products are evaluated. Although this is very useful and a recommended approach in the pre-design phase, the **Ergodesign Methodology for Product Design** requires the participation of users in the further steps of the design process, not just "consulting" them, but also "involving" them in the entire process in a participatory way. This is the essence of co-design approach.

Some crucial aspects, such as task analysis, user trials, model, and prototype evaluations, require the constant involvement of users. Focus groups, as a technique of a consultative nature, do not show how users behave with products. Because of this, it is recommended to select a group of about eight product direct users and two indirect users from the focus group sessions to assist in the following phases of the design process. This group of participants will be called the **User Panel** henceforth.

Observing the participants in the focus group sessions should be a good way to select those candidates who make a good contribution to take part in the **User Panel**. Some factors such as the participants' capacity for criticism, observation, enthusiasm, and giving useful suggestions; the guarantee of the availability of the participants in the several **User Panel** meetings, in terms of time commitment; and a suitable financial remuneration should be taken into account.

Although it can be argued that users are in general non-technical, have a lack of knowledge of how the product works and the adequacy of different materials and components, and know very little about the limitations imposed by the manufacturing process; their participation in the design process can be justified by the fact that their unique experience using the product can be transformed into a rich, creative, and innovative source of information which will improve the quality and usability of the product.

According to Feeney (1996), many of the practical problems in involving users in the design and manufacturing process arise as a result of traditional attitudes on the part of the manufacturers and designers rather than problems to be overcome.

To improve the communication between the **User Panel** and the design team, participants should be informed about how products are designed, manufactured, and sold, including the constraints imposed by the production process. This procedure will start the involvement of the **User Panel**, stimulating them to question the way things have been done and preparing them to propose new and creative ideas and solutions.

A number of sessions should be established at significant points of the design process to enable the participation of the **User Panel**, with the design team, having a discussion and making decisions related to issues to be incorporated in the future steps of the design process. At each design review, the panel should be informed about the design development, asked to discuss the results, and point out suggestions for the further phases. The involvement of the manufacturing, marketing, and commercial personnel in the design review sessions will contribute to giving extra inputs to the discussion and may contribute to improving the quality of the results.

A chairperson, maybe one of the directors of the company, should be chosen to run the **User Panel** meetings. It is important to take into account that some review decisions may not have the full agreement of the **User Panel**. Wherever possible, this must be avoided through discussion or modifications to the design to ensure that the needs and wishes of all members of the panel are met. If these are not possible, a decision might be taken by the Chairperson.

In addition to taking part in the **User Panels** and contributing to resolving design conflicts, the participation of the product direct and indirect users in the phase of **Testing and Verification** is essential. Practical participation of the members of the panel includes task analysis, user trials, evaluation of mock-ups, real and digital models, and prototypes and instruction manual design. Eventually, the **User Panel** can be supplemented by other subjects recruited for special tests, e.g., anthropometric tests that require subjects of specific body size.

(b) Defining the Intervention Phases of the User Panel

An indication of each phase of the **Ergodesign Methodology for Product Design** with the major stakeholders involved in the design process, including the participation of the **User Panel**, is given in Table 7.1. The participation of member(s) of the

TABLE 7.1

The Design Phases Involve Several Stakeholders in the Design Process

	Stakeholders			
Production Phases	User Panel	Design Personnel	Marketing Personnel	Management and/or other Personnel
Preliminary Strategic Planning		•	•	•
• Define a business plan and overall budget				•
• Lay down outline timetables				•
• Establish preliminary guidelines for innovation		•	•	•
• Define applicable technologies		•		•
• Define the target market			•	
• Identify competitive products		•	•	
Approaching the Users		•	•	
• Design focus group		•[a]	•[a]	
• Carry out focus group sessions		•[a]	•[a]	
• Result of the focus group including the establishment of user needs		•[a]	•[a]	
• Select users for the *User Panel*		•	•	
Investigating the Problem	•	•		
• Recognizing the Problem		•		
• Delimitation of the Problem	•	•		
• Formulating the Problem		•		
Product Planning	•	•		
• Carrying out Task analysis	•	•		
• Refining user needs	•	•		
• Reviewing the state of the art		•		
• Applying QFD to product development		•		
• Elaborating the *Product Design Specification* document		•		
Design Creation	•	•	•	
• Generating concepts		•		
• Evaluating concepts	•	•		
• Selecting concepts	•	•		
Detailing concepts				
• Refining concepts		•		
• Detailing design		•		
• Designing the user manual		•		
• Designing promotional material		•	•	
Prototyping		•		
• Building the prototype		•		
Testing and Verification	•	•	•	
• Mock-ups and models evaluation	•	•		

(Continued)

TABLE 7.1 (CONTINUED)
The Design Phases Involve Several Stakeholders in the Design Process

	Stakeholders			
Production Phases	User Panel	Design Personnel	Marketing Personnel	Management and/or other Personnel
• Prototype evaluation	•	•		
• Prototype modification		•		
• Re-testing prototype	•	•		
• Test and review user manual	•	•		
• Test and review promotional material	•	•	•	
Product Production			•	•
• Production development				•
• Production planning			•	•
Manufacturing and Assembly				•
Product Marketing			•	•
Customer Support			•	•

^a An external consultant can be hired to run the focus group instead of the design and/or marketing team.

design and/or marketing team in running the focus group sessions may be substituted by a qualified external consultant. The phases of **Product Production**, **Product Marketing**, and **Customer Support** are not described in detail because it does not involve the direct participation of the designer.

7.2.3 INVESTIGATING THE PROBLEM

The definition of the steps included in the design process depends strongly on the correct identification of the problems to be solved. Broadly speaking, products in general, including wheelchairs, can be understood as material systems comprised of a set of properties. These properties are made to fulfill functions that, in their turn, will permit users to perform specific actions that will or will not meet their needs. Figure 7.2 presents a model of the **Human System Material Interaction**. In this model, products interact with humans based on their properties and functions. This will cause the user to perform actions on the product. As a result, the user will be satisfied or dissatisfied. If the product does not fulfill user needs entirely (dissatisfaction), a potential situation for redesigning the product or designing a new one can be established to overcome the identified problems. We will use wheelchairs to exemplify this phase of the methodology.

The problems to be investigated are obtained from the needs of the users that were previously identified in the **Focus Group** sessions.

Thus, investigating the problem will give the design team basis to decide what to do and how to do it. This stage of the methodology is based on Moraes and Mont'Alvão (2010).

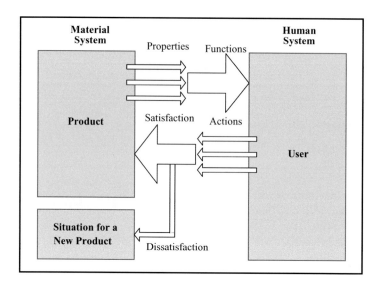

FIGURE 7.2   Human system material interaction. Source: The Author.

ATTENTION

According to Moraes and Mont'Alvão (2010), the stage of **Investigating the problem or Problematization** can be divided into three phases:

- Recognizing the problem;
- Delimiting the problem; and
- Formulating the problem.

7.2.3.1   Recognizing the Problem

The approach to investigating a problematic situation can be carried out by describing what is lacking in the product or situation to be analyzed, in terms of fulfilling user needs, and/or what exists but actually does not perform its actions as required to meet user needs and promote satisfaction.

Recognizing the problem corresponds to investigating the most serious and flagrant problems that immediately appear in the analysis of the problematic situation. This preliminary phase provides the first inputs to the design team and establishes an initial set of problems that needs deep investigation. These inputs are presented in the form of a **List of Problems** collected in a non-systematic way.

Undoubtedly, focus groups are an excellent tool for recognizing the problems in this initial phase. Certainly, product users will be able to verbalize their wishes, wants, and complaints concerning the products they currently use and other products that may be introduced.

7.2.3.2   Delimiting the Problem

The previous step generates a **List of Problems** obtained from the company and the focus group sessions and non-systematically identified. The **List of Problems** should be now selected, classified, and expanded, emphasizing the most relevant design problems. This should be done by the design team and further submitted for appreciation and comment by the **User Panel**.

The selection and classification of the various aspects of the problem situation are described through the **Analysis of Dysfunction of the Product–User Interface**. These dysfunctions must be identified as problems that affect the user interface with the product or system. Examples of the several distinct dysfunctions applicable to the product–user interface (wheelchair) are shown in Tables 7.2, 7.3, and 7.4 (adapted from Moraes and Mont'Alvão, 2010; Moraes, 1992 and Soares, 1990).

ATTENTION

User–Product Interface Dysfunctions can be divided into three types and are composed of several problems associated with each type of dysfunction (Moraes and Mont'Alvão, 2010; Moraes, 1992 and Soares, 1990). The following are examples of problems associated with each of the dysfunctions:

- **Ergonomic dysfunctions** – Interface, postural, dimensional, instrumental, informational, perceptual, control, communicational, operational, cognitive, movement, natural, and accidental problems.
- **Human dysfunctions** – Sensory-physiological (vision, hearing, touch) and psychoneurophysiological problems.
- **Machine dysfunctions** – Structural and moving, resistance and productivity, socio-cultural, and semiological and sensory-formal problems.

It is important to call attention to the fact that, although these pointed out problems are related to wheelchairs, many of them can be applied to any other consumer product, and, for this, they must be adapted to the product under analysis. Also, the designer must identify each or other categories of problems when carrying out his or her particular analysis of the dysfunction of the product user interface.

ATTENTION

Each of the problems identified in the **User–Product Interface Dysfunction** stage must be properly analyzed and represented through the **Problem Formulation Table**, as described in the following step.

This phase of **Analysis of Dysfunction of the Product–User Interface** will be more successful with the use of **photographs** to illustrate the problems. It can be

TABLE 7.2

Example of Problems with Ergonomic Dysfunctions in the Product–User Interface for the Wheelchair Product

Ergonomics Dysfunction Problems

Human–product interface problems

- Poor posture as a result of the inadequate location of controls that respond to actions by wheelchair users or carers.
- Use of inadequate anthropometric values in defining wheelchair dimensions.
- Location of displays out of the field of vision of extreme users.
- Location of controls outside the users' dynamic area of comfort.
- Limited space to accommodate trunk and legs.
- Poor support to accommodate arms and feet.

Instrumental problems

- Displays and/or controls are provided with no consideration of prioritization, ordering, and standardization.
- Movements of displays and/or controls with no consideration of stereotypes of movement and consistency.

Informational problems

- Poor location of objects to be perceived and discriminated as a function of their shape and distance from the user.
- Poor visibility of warning and graphic signals.
- Poor legibility of characters.

Control problems

- Pain in fingers, wrists, elbows, arms, shoulders, trunk, feet, and legs causing by repetitive effort, resistance or vibration of the controls, and poor position of the hands in consequence of position and movements of the manual controls.
- Poor dimension and shape of the manual controls exerting pressure on specific points of the hands.
- Poor dimension and shape of foot support.
- Lack of safety in controls with possibilities of electrical shocks, burning, cutting, or injuries.
- Location of handles and foot supports out of the user's dynamic reach area and biomechanic angles of comfort.
- Difficulty to visualize and/or reach components that require maintenance and repair.

Cognitive problems

- Deficiency in the operational logic resulting from the layout and movement of displays and controls, without considering the consistency of the system, users' stereotypes, and operational images.
- Poor comprehensibility of the graphic symbols as a result of cultural incompatibility, newness, or lack of knowledge of the codes used.

Movement problems

- Excessive weight of the wheelchair, causing difficulties in lifting for storage into car boots or pushing on difficult terrain.
- Excessive weight of some wheelchair components that need to be replaced (e.g., battery).
- Wheel sizes incompatible with certain terrain.

Natural problems

- Lack of accessories to protect the user against rain, snow, or cold weather.

Instructional problems

- Poor quality of instructional manuals.

TABLE 7.3

Example of Human Dysfunction Problems in the Product–User Interface for the Wheelchair Product

Human Dysfunction Problems

Postural problems
- Back pain resulting from poor posture assumed when activating controls and other components, getting visual information from displays (wheelchair users), or using push handles at inappropriate height (carers).
- Damage to the spinal column due to lifting and transporting excessively heavy wheelchairs.
- Muscle fatigue resulting from repetitive efforts and poor posture when users and carers are pushing the wheelchair.

Social Problems
- Difficulty in being socially active due to using an unfriendly wheelchair, which is incompatible with use in tight public places such as shops and pubs.
- Lack of self-esteem due to the use of a product with a design that reinforces the image of disability.

---

done using members of the **User Panel** and will be a strong tool to persuade the company to invest in a new product.

7.2.3.3 Formulating the Problem

In this last phase of investigating the problem, the situation is reduced to its most significant and soluble aspects, considering the competence of the product development team, the available knowledge, and what was required by the users and the company. This can be shown by using a table named **Formulation of Problems**.

The table should contain the following items:

- The main **problems** encountered in the **User–Product Interface Dysfunction** stage,
- The **requirements (needs)** for a new design generated from the problems encountered,
- **Human (physiological) problems** that may affect the user due to the problems encountered,
- **Human costs** (pain or discomfort) that can affect the user due to the problem experienced,
- **Suggestions** for improvements to the new design, and
- Any **restrictions** that may occur to prevent the new design from being adopted.

For the purpose of illustration, Table 7.5 shows some possible examples of ergonomic dysfunction problems that may be contained in the **Table of Formulation of Problems** using the example of the design of wheelchairs. This table aims to

TABLE 7.4

Example of Machine Malfunction Problems in the Product–User Interface for the Wheelchair Product

Machine Dysfunction Problems

Structural and movement problems

- Poor stability of the wheelchair structure.
- Too little or too great resistance to effort.
- Poor adjustability and interchangeability of components.
- Sharp edges and protruding nuts and bolts in the wheelchair structure.
- Frame difficult to unfold.
- Lack of security for fixed wheelchair components.
- The movement system is excessively stiff or flexible.
- Noise in the movement system.
- Failure in the brake system.
- Lack of flexibility in the use of accessories.

Problems of poor performance of components and subsystems

- Lack of confidence due to failures and non-functioning of components.
- Components functioning below the standard required.
- Poor durability of subsystems and components.

Problems of robustness, reliability, standardization, and manufacture

- Lack of resistance of materials to bad weather.
- Lack of resistance of materials to physical aggression.
- Lack of standardization, modularization, and interchangeability of components with consequences for the cost of the product, speed of production, and the achievement of high levels of quality during manufacture.
- Costs of manufacturing incompatible with the scale of production.
- Excessive use of different materials with an increase in the number of manufacturing operations and costs.
- Inadequacy of the manufacturing process to the company capacity in terms of raw material and equipment available, know-how, and qualified personnel.

Social-cultural and semiology problems

- Design inadequacy in terms of representation of the user's uniqueness, values, and status.

Esthetic/form problems

- Lack of originality in the wheelchair design with no distinction between the company's and competitor's products.
- Design esthetically unpleasant with poor configuration and inadequate use of materials, colors, and textures.

Technological problems

- Technology unknown to the user without references and knowledge repertoire proper to operate.

guide the design for the next steps of the design methodology working as a verbal modeling of the product to be designed. It should be noted that the column "Suggestions" meets the needs presented in the column "Design Requirements". Observe that this table comprises only a few examples of problems found in the analysis.

TABLE 7.5

Some Examples of Ergonomic Dysfunction Problems for a Table of Formulation of Problems

Problems	Design Requirements	Human Problems	Human costs	Suggestions	Design constraints
Interface					
The backrest does not support the lower back	Backrest profile which considers the buttock protrusion and supports the lumbar region	Dorsal kyphosis and flattening of the lumbar curve	Pain in the back	Provide a new backrest profile	Available technology Lack of interest of buyers and manufacturers
Inappropriate support to accommodate the feet	Considers the length of the feet of the biggest users Considers the height of the legs of smaller and bigger users	Legs do not touch the foot support Pressure in the posterior keen region (popliteal region)	Discomfort	Provide an adjustable foot support	Lack of interest of buyers and manufacturers
Inappropriate location of push handles	Considers the height of the elbow of the biggest and smaller carers and defines the height of push handles	Flexion of the lumbar spine	Pain in the lower back Pain in the neck	Provide adjustable push handles	Lack of interest of buyers and manufacturers
Control					
Inappropriate shape of the hand controls	Profile that does not cause pressure on the user's hands	Pressure on specific areas of the hands Ulnar/radial deviation	Pain in hands and wrist	Provide new profile for the hand controls	Available technology Lack of interest of buyers and manufacturers

ATTENTION

The data contained in the **Problem Formulation Table** should be used to compose the **Design Requirements** for the new product.

7.2.4 PRODUCT PLANNING

The problems are now adequately defined, and the project boundaries are established. This step involves finding the information directly relevant to the designers' further activities of generating and selecting feasible solutions to the creation of new product models.

ATTENTION

The **Product Planning** phase covers the following steps:

- Carrying out a task analysis.
- Refining user needs.
- Reviewing the existing state of the art.
- Formulating the design specifications.

7.2.4.1 Carrying Out a Task Analysis

The **Task Analysis** must be performed with one or more products similar to the one intended to be designed in order to understand the user's interactions with the product. **Task analysis** is an important method in the ergonomics repertoire. It is a method for producing a hierarchical flowchart of all the things the user will do with a product called **Task Activity Flowchart**.

Each task in which the user engages using the product can be divided into a set of activities. Each of these activities can be then broken down further into sub-activities; sub-activities can generally be further divided into sub-subactivities, and so on. Tasks and activities can be organized in terms of product subsystems and subsystems. In this way, the **Task Analysis** comprises the hierarchical and sequential presentation of the activities performed when the user uses a product, or a worker performs activities at the workplace. This technique is very familiar to all ergonomists. An extensive review of **Task Analysis** can be found in Annet and Stanton (2000).

ATTENTION

Task analysis provides the designer and ergonomist with details on:

- The sequence in which the user uses the product.
- The place in the hierarchy of each activity.

- User–product interface requirements.
- Product evaluation and decisions that must be made in the design.
- Task times and environmental conditions.

In using the **Ergodesign Methodology for Product Design**, designers and ergonomists should use **Task Analysis** as a tool to examine the product–user interface in detail. Although this technique is usually carried out by ergonomists, the participation of the designers is essential because task analysis provides a rich source of inspiration on new product concepts and a rational basis for design decisions. **Task Analysis** will also provide useful information about anthropometric aspects of users during the use of some products.

The use of **Task Analysis** in the **Ergodesign Methodology for Product Design** should complement the data obtained from the **Analysis of Dysfunction of the Product–User Interface** described in the previous phase of **Delimitation of the Problem**. As product users, the members of the **User Panel** should be invited to take part in the **Task Analysis** for investigation of both the tasks performed by the indirect users and those performed by the direct users themselves.

Details of various **Task Analysis** approaches and illustrations of operational sequence diagrams, useful for product design, can be found in a number of books and book chapters, including those by Stuster (2019); Annet and Stanton (2000); Baxter (1995); Chapanis (1996); Cushman and Rosenberg (1991); Dumas and Redish (1999); and Kirwan and Ainsworth (1993).

ATTENTION

According to Moraes (1992), to understand the demands of the task, during Task Analysis, four elements must be investigated:

- The content of the task.
- Its implications in terms of taking information and visual requirements.
- What they determine in terms of actional responses such as manipulation of commands, levers, triggers, etc.
- What they determine in terms of manipulation of commands such as levers, triggers, etc.

As an example of a **Task Analysis**, we chose an activity more complex than the handling of a wheelchair: the task of a user (seamstress) using a product (sewing machine) in the clothing industry. This example was part of a consultancy work carried out by the author. The task is represented by a **Task Activity Flowchart** for sewing a shirt cuff (Figure 7.3). The task analysis is described as following:

- System: sewing a men's shirt.
- Sub-system: sewing the cuff of a men's shirt.
- Task: a seamstress, using a sewing machine in a clothing industry.

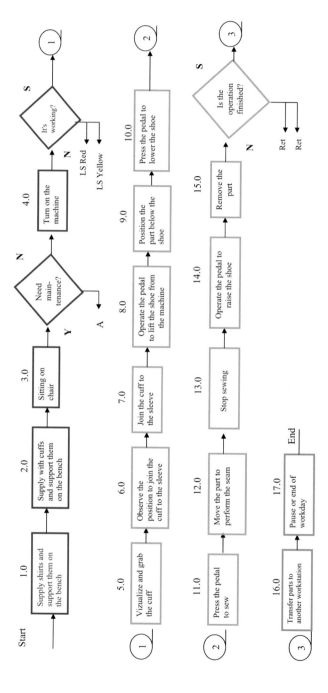

FIGURE 7.3 Flowchart of the activities of the task of sewing a shirt cuff. Source: The Author.

- Activity: is defined as each of the components of the task. For example: taking information, operating pedals, moving sewn parts, communicating with colleagues.

The sewing task example can be used as a reference for the Task Analysis of any consumer product.

In the **Task Analysis** presented in the example, it can be seen that all activities are serial (they occur in sequence, one after the other), but, in some situations, which are not covered in this example, the tasks can be simultaneous or alternating. They are represented by rectangular boxes (to express regular activities) or a diamond (to express decision making). In this example, there are two light signals (LS) on the workstation: one red and one yellow. These signs mean that one of the sewing machines has broken and that it has no other to replace (red) and that the machine has broken and another machine in the production cell to replace (yellow). Thus, it can be seen that details can be added to the **Task Activities Flowchart** depending on the particulars of the various activities.

The flowchart can be represented by color codes corresponding to activities, for example, initialization, ordinary, completion, and maintenance.

- **Start-up activities**, shown in blue, including all activities before sewing starts, e.g., filling and starting the machine, searching for material, etc.
- **Ordinary activities**, represented in orange, comprise the sewing activities themselves.
- **Finishing activities**, shown in green, which correspond to the activities' closing stage, e.g. transfer parts to another station, turn off the machine, etc.
- **Maintenance activities**, not shown in this example, means a break from ordinary work activities to supply new parts or change the line.

7.2.4.2   Reviewing the Existing State of the Art

Reviewing the existing state of the art is paramount in determining the commercial success of the product to be launched in the marketplace. Apart from collecting information on the ergonomics, technical product specification, and safety and regulatory standards, information on competing products must be gathered in order to specify the position of existing products, both the companies and competitors.

(a) Review of Literature

A literature search will enable the design team to find technical reports, books, magazines, journals, and conference proceedings with relevant papers related to issues of the product to be designed. Steps recommended to the designer in finding information about the product issues include:

- Examining lists of references that follow articles on the product and related topics in journals, conference proceedings, and books
- Reading newsletters of professional societies and technical groups
- Obtaining listings and abstracts of recent reports prepared under government contracts
- Searching computer databases and the internet
- Searching patents and standards.

Information from the literature on ergonomics related to problems associated with people with disabilities in general and wheelchairs, in particular, is very rare. Anthropometric data available in the literature to define the dimensions and body shapes of people with disabilities are also overwhelming.

It is recommended, as a way to overcome this lack of anthropometric data, that designers carry out their tests using members of the **User Panel** that can represent the extremes of the population: for example, the short, the tall, the thin, the fat. Performing user task analysis and testing with members of the **User Panel** will also assist in establishing users' anthropometric dimensions. We recommend taking care when collecting anthropometric data as they need to follow a scientific procedure.

Although designers are destined, at least for the moment, to fail to search for comprehensive information about the dimensions of people in their wheelchairs, the ergonomics literature has several other extremely useful data sources for product design. These sources include data on user behavior, physical and mental data, techniques for applying task analysis, questions related to user testing and experimentation, data on displays, information design, controls, and product safety and control arrangements.

NOTE

Some books on ergonomic literature that are useful for product design include Kroemer et al. (2018), Kroemer (2017), Soares and Rebelo (2017), Bridger (2017), Shorrock and Williams (2016), Tillman et al. (2016), Wheelchair Foundation (2016), Wilson and Sharples (2015), Salvendy (2012), Karwowski et al. (2011a, b), Green and Jordan (1999), Jordan (1998a), Roebuck et al. (1995), Sanders and McCormick (1993), and Cushman and Rosenberg (1991).

(b) Technical Standards

Technical standards, mandatory or voluntary, contribute to improving the quality and, above all, the safety of products. The standards for equipment for people with disabilities are listed and indexed in the standards of ISO/TC 173: Assistive products (2019). In fact, manufacturers must comply with ISO standards in order to export their products to other countries. Designers should check if there are any rules or regulations applied to the consumer markets where the products will be distributed. This procedure aims to ensure compliance with the requirements from the beginning of the design phases.

7.2.4.3 Refining User Needs

The previous phases of the design process identified a considerable number of needs for users of the product. Returning to the example of wheelchair users, the needs of these users were identified as a result of the expression of wishes, desires, and complaints. The needs verbalized by the users are expressed in their own language and, although they are a clear expression of their interests and desires, they are not specifically described as guiding to designing and engineering the product. This leaves designers and engineers with the task of interpreting the users' needs with a minimum of subjective interpretation.

The most appropriate way to translate the user needs into the design process is to establish the **Product Requirements**. They comprise a set of specifications that will address what the product has to do, in a precise and measurable way, to meet user needs. This means that, for example, a user needs to "reduce the weight of wheelchair", will correspond to the specification that "the weight of the wheelchair should be 10 kg". Ideally, each user need should correspond to only one value specification, although this is frequently not possible. Issues related to how the product will perform to satisfy user needs are not yet addressed in this phase of the methodology.

User needs, previously established in the form of a list, arising from the **Problem Formulation Table**, have now to be selected (refined), categorized, and ascribed a level of importance. The selection of the user needs must include those who are within the designer's competence to solve. Categorizing them refers to associating each identified and selected need with the respective subsystem of the wheelchair. For instance: "reduce the weight of wheelchair", "produce wheelchair foldable", "reduce vibration in the handles", "allow ease of traversal of difficult terrain", are needs that can be associated with the subsystem "Structure". "Seat and backrest should be made to stop stretching and sagging", "provide further cushion on seat and backrest", "washable and easily removable upholstery", are examples of needs related to the subsystem "Seat-backrest". The designer should consult the **User Panel** to attribute levels of importance to each need varying from 1 (the less important need) to 5 (the most important need). This will be essential to decide which user needs have to be taken in resolving subsequent design dilemmas.

ATTENTION

The **Refining User Needs** step will define the **User Requirements** for the new product design and should be presented through a **Refined User Needs List**.

The **Refined User Needs List** means that each need identified in the **User Requirements** stage must be associated with a corresponding metric. Attention should be paid to the fact that there will be some needs that cannot be easily transformed into quantifiable metrics. In this case, the user's need must be maintained, and the metric is evaluated as "subjective". Table 7.6 shows an example of user needs for the "Structure" subsystem of the wheelchair, its relative importance, associated metrics, and units of measurement.

To provide a metric value corresponding to each need is necessary to guarantee that meeting the specification will lead to satisfaction of the associated user needs. Tables such as 7.6 will be a key element in the **House of Quality**. **Refined User Needs List** is one of the components that will comprise part of the QFD matrix, which is described later.

TABLE 7.6
List of Refined User Needs and their Associated Metrics

No.	Subsystem	Need	Imp.	Metrics	Units
1	Structure	Reduce the weight of the wheelchair	5	Total mass	kg
2	Structure	Produce foldable wheelchair	4	Fold width	mm
3	Structure	Reduce vibration in the handles	3	Attenuation from push bar to the main structure at 10 Hz	dB
4	Structure	Allow easy traversal of difficult terrain	4	Spring preload	N
5	Structure	Easy to remove wheels	1	Time to disassemble/assemble	min
6	Structure	A wide variety of wheels and tires fit the wheelchair	2	Headset size Steer tube diameter Wheel sizes Castor sizes Maximum tire width	mm mm mm mm mm
7	Structure	Easy access to maintenance of the components	2	Time to disassemble/assemble	min
8	Structure	Sharp edges are smoothed off	3	Sharp edges are smoothed off	subj
9	Structure	Easy to fit accessories	3	Time to assemble the accessories	min
10	Structure	Easy to maneuver	4	Minimum corridor width of 1000 mm	mm
11	Structure	Last a long time	4	Test of steer tube duration	hours
12	Structure	Provide good stability	5	Test of stability ISO 7176-1	degree
13	Structure	Easy of curb climbing	3	Test of obstacle climbing ability ISO 7176-10	mm
14	Structure	Is safe?	5	Fatigue test	N and curb drops
15	Structure	Is not expensive	5	Unit manufacturing costs	£

7.2.4.4 Formulating the Design Specifications

This stage will consist of the **Analysis and Evaluation of Competing Products** and the application of the **Quality Function Deployment Matrix**.

(a) Analysis and Evaluation of Competitive Products

Analyzing and evaluating competitive products is an absolutely essential activity to determine the strengths and weaknesses of competing products in relation to the company's own product. Information gathering from competitive products will clarify problems associated with existing products, which must be overcome to increase the chances of success for the company's own new product. This phase will generate a **Table of Competitive Products.** This table should be divided in two: the first based on **Metrics**, and the second based on **User Satisfaction**. Both will be part of the QFD matrix later.

A database is the most effective way to store and retrieve information on the characteristics of competitive products. In this way, data can be easily updated and used and can provide on-hand information about opportunities for product improvement. Examples of information to be included in the database are:

- The results of ergonomics tests.
- Findings from the direct observation of product use.
- The results of surveys and interviews with product direct and indirect users.
- The outcomes of evaluations by experts (marketing, engineering, ergonomics, industrial design).
- Product reviews in consumer publications, trade publications, and design publications.
- Product description in sales literature and advertising.

Table of Competitive Products Based on Metrics

Gathering data from competitor products is very time-consuming and may involve purchasing, testing, disassembling, and estimating their production costs. Independent evaluations, such as those contained in Consumer Reports magazine (www.consumerreports.org), from the UK Consumers Association and reports from The Medicines and Healthcare Products Regulatory Agency (MHRA), UK, may be a good source of obtaining data. It is important to draw attention to the fact that sometimes the data included in competitors' catalogs and supporting literature are not accurate.

An example of a **Table of Competitive Products Based on Metrics** is shown in Table 7.7. Note that some data has been identified previously in Table 7.6 and will be properly transported to Table 7.7. The contents of each column in the table are:
Metric number as previously adopted in Table 7.6,

The **Metric number** (which will be the same as the metric number and will be useful to be used in the next table),

- The **Metrics** (measures referring to each need, as identified in Table 7.6),
- The **Importance** of this need as defined by users,
- The **unit** adopted for each metric and
- The **measures** found for each analyzed product from competing companies.

The data in the table are fictitious, with companies identified by letters. The only purpose of Table 7.7 is to illustrate the technique used in this methodology.

Table of Competitive Products Based on User Satisfaction

Table 7.8 (**Table of competing wheelchairs based on user satisfaction**) shows a comparison amongst competing wheelchairs based on users' perceived satisfaction

TABLE 7.7

Table of Competitive Wheelchairs Based on Metrics

Metric No.	Need Nos.	Metrics	Imp.	Units	A	B	C	D	E
							Companies		
1	1	Total mass	5	kg	15.5	20.0	17.3	16.8	18.0
2	2	Fold width	4	mm	330	580	910	730	815
3	3	Attenuation from push bar to the main structure at 10 Hz	3	dB	12	15	14	12	15
4	4	Preload on the suspension spring	4	N	480	760	500	520	680
5	5	Time to disassemble/ assemble wheels	1	min/ sec	15m 18s	38m 40s	27m 45s	32m 20s	35m 55s
6	6	Headset sizes	2	mm	1.000 1.125	1.000 1.000	1.000 1.250	1.125 1.250	1.125 1.250
7	6	Steer tube diameter	2	mm	254	254	254	254	254
8	6	Wheel size	2	mm	609	558	609	508	628
9	6	Castor size	2	mm	127	190	190	127	190
10	6	Maximum tire width	2	mm	38	44	44	44	44
11	7	Time to disassemble/ assemble components	2	min/ sec	8m 42s	10m 15s	12m 10s	9m 23s	10m 45s
12	8	Sharp edges smoothed off	3	subj	4	3	5	3	2
13	9	Time to assemble the accessories	3	min/ sec	3m 12s	5m 23s	4m 14s	3m 45s	6m 23s
14	10	Minimum corridor width of 1000 mm	4	mm	1125	1450	1350	1500	1400
15	11	Test of steer tube duration	4	hours					

(Continued)

TABLE 7.7 (CONTINUED)
Table of Competitive Wheelchairs Based on Metrics

Metric No.	Need Nos.	Metrics	Imp.	Units	Companies				
					A	B	C	D	E
16	12	Test of stability ISO 7176-1 (using a 100kg test dummy)	5	degree	14°	15°	16°	17°	15°
					>20°	>20°	>18°	>20°	>18°
					>20°	>20°	>18°	>20°	>18°
17	13	Test of obstacle climbing ability ISO 7176-10	3	mm	20	25	23	25	25
18	14	Fatigue test	5	N	1000	1230	1350	1420	1350
				curb drops	6248	8453	10450	9821	10115
19	15	Unit manufacturing costs	5	£	1675	1954	1825	2200	2650

TABLE 7.8

Table of Competing Wheelchairs Based on User Satisfaction

Needs No.	Needs	Imp.	Companies A	B	C	D	E
1	Reduce the weight of the wheelchair	5	4	1	3	1	2
2	Produce foldable wheelchair	4	3	2	1	1	1
3	Reduce vibration in the handles	3	2	1	2	2	1
4	Allow easy traversal of difficult terrain	4	1	2	3	1	1
5	Easy to remove wheels	1	3	2	1	1	2
6	A wide variety of wheels and tires fit the wheelchair	2	1	2	3	1	2
7	Easy access for maintenance of the components	2	3	2	1	3	2
8	Sharp edges smoothed off	3	2	1	4	2	1
9	Easy to fit accessories	3	3	2	3	2	2
10	Easy to maneuver	4	4	2	3	2	1
11	Lasts a long time	4	2	3	3	4	4
12	Provides good stability	5	1	3	3	4	4
13	Ease of curb climbing	3	2	2	3	3	4
14	Is safe	5	3	3	4	4	4
15	Is not expensive	5	3	2	1	1	1

with the degree to which the different wheelchairs satisfy their needs. The satisfaction level is presented on a scale (rank) from 1 to 5 (where 1 is not very important and 5 is very important). Wheelchairs that scored more correspond to greater perceived satisfaction of the user needs. This is a subjective evaluation and should be carried out with the assistance of the **User Panel**.

(b) Applying Quality Function Deployment to Product Development

The appropriateness of **Quality Function Deployment** (QFD) as a formalized method of matching the expressed needs of the users to the features and functions of the product makes it an ideal choice to be used as part of a human-centered method for product design. Quality function deployment was reviewed in Chapter 6, Subchapter 6.2.

As discussed in Chapter 6, the **House of Quality** (HOQ) is a multidimensional figure that shows the relationship of the user needs to the metric characteristics of the product. Figure 7.4 shows a partially completed QFD figure for the design of a wheelchair (adapted from Maritan, 2015; Akao, 2004; Hauser and Clausing, 1988; Menon et al., 1994; Pugh, 1991 and Sullivan, 1986). The data are fictional and were used only as an example of the application of this technique to the development of products, in this case, wheelchairs.

According to Akao (2004), QFD will convert consumer needs (requirements) into product engineering characteristics (metrics). This will be done by deployment the

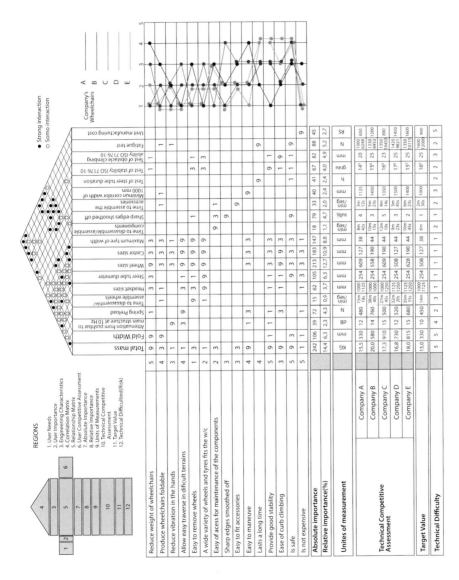

FIGURE 7.4 Example of a partially completed QFD table for the design of a wheelchair. Source: The Author.

relationships between the needs identified by the user and the product characteristics through the various houses of quality. According to the author, these relationships occur at three levels: 1) Extraction, 2) Correlation, and 3) Conversion.

- **Extraction**: This is the process of creating a table from another, that is, transporting data from one table to be used in another.
- **Correlation**: Is the process of identifying the intensity of the relationship between the data from two tables that make up the matrix.
- **Conversion**: There are two types of conversions: 1) **Quantitative conversion**, when the original data is transformed from the verbalizations collected in the surveys with the users into converted data (metrics) to be used as requirements (needs) of the users; 2) **Qualitative conversion**, when relative importance data (attributed weights) is transferred to the data of another table, considering the intensity of the relationship existing between them.

The HOQ consists of 12 regions. These are shown in brief form in the right-hand corner of Figure 7.4. Each of these regions is described below, and appropriate sources of information to elaborate on each region are also given. We present the QFD regions as follows.

1. **User Requirements/User Needs** (Region 1) are grouped by arranging them according to each subsystem. An illustration of **User Needs** for the structure subsystem is given in column 3 (**Need**), extracted from Table 7.6.
2. **User Importance** (Region 2) are ratings or weighted values, as indicated by the users. This is illustrated in column 4 (**Imp**), extracted from Table 7.6.
3. **Engineering** or **Metric Characteristics** (Region 3) are established in terms of measurable quantities (**Metrics**), extracted from Table 7.6, column 5.
4. **Correlation Matrix** (Region 4) shows the relationship between the different product's engineering characteristics, in this case, wheelchair. It presents the interrelations between the **Engineering Characteristics** as **Strong Interaction** or **Some Interaction**. This is illustrated in the triangular roof of the HOQ shown in Figure 7.4.
5. A **Relationship Matrix** (Region 5) identifying the levels of influence and effect between each engineering characteristic and the users' requirements (needs). An arbitrary scale of 9 (strong), 3 (moderate), and 1 (weak) is used to weigh those engineering characteristics that affect user needs. The relationship matrix is shown in the body of the HOQ in Figure 7.4, where the relevant values are 1, 3, or 9. These scale values aid in the definition of issues of the highest absolute importance, as described in 8 below.
6. The **User Competitive Assessment** (Region 6) is a summary of a 5-point scale (higher value is better) of the extent to which a company's wheelchair (A, B, C, D, or E) meets user' needs. The higher the value, the more it will meet the corresponding requirement. The summary is given as a graphical profile on the right-hand side of Figure 7.4. These data are a direct transcription of values on user satisfaction given, extracted from Table 7.8.

7. **Absolute Importance** (Region 7) of each metric in the QFD Table is found by adding the product of the numerical sum of each element in the columns of the **Relationship Matrix** (the body of the House of Quality) multiplied by the value assigned by the user for each importance ranking. For instance, in the first cell of the column "Total mass", the value 9 is multiplied by the user importance value 5, giving a total of 45. This operation is repeated in the entire column and, with the sum of the total values, an absolute value of 242 is obtained. In the next column, "Fold width", the same value 9 was assigned. In turn, it was also multiplied by the value of the importance given by the user, which was 5, resulting in a total of 45. Performing the sums of all the operations gives an absolute value of 106, and so on. An absolute importance row is shown in the HOQ (Figure 7.4).

8. **Relative importance** (Region 8) is the determination of the percentage obtained from the total numerical score for each engineering characteristic multiplied by one hundred and the sum of all metrics. The total numerical score is the sum of all the values of absolute importance appearing in region 7 (in this sample, the total numerical score obtained from the sum of the engineering characteristic is 1677). The relative importance percentage for each engineering characteristic value is obtained by multiplying the total numerical of each metric by 100 (e.g., 242 in the first column) and dividing it by the total numerical score. For instance, for the column "Total mass": $242 \times 100 = 24200 \div 1677 = 14.4$). For the "Fold Width" column: $106 \times 100 = 10600 \div 1677 = 6.3$. Those engineering characteristic with the highest ranking are the characteristics relating to a requirement considered the most important to the user and should be prioritized by the design team.

9. **Units of measurement** (Region 9) are the values corresponding to each engineering characteristic (e.g., hour, minute, kilograms, millimeters, etc.). These units are given toward the foot of the HOQ (Figure 7.4) and are extracted from column 5 in Table 7.7.

10. **Technical Competitive Assessment** (Region 10) compares the competitor's specifications for each of the product's engineering characteristic and the proposed specification to either meet or exceed each characteristic (the data are those shown in Table 7.7 and are transcribed into Figure 7.4).

11. **Target values** (Region 11) are determined for each of the wheelchair's engineering characteristic. These values are frequently determined, in part, from benchmarking data and from an independent assessment of how strongly the values impact the product's performance, attributes, and features.

12. **Technical difficulty** (Region 12) is a judgment, on a scale from 1 to 5, based on the experience of the design team and indicates the ease with which each of the product's specifications can be achieved. The lower the number, the easier it is and, consequently, the risk of not meeting that characteristic is lower.

The result of this stage of the ergodesign methodology is the **Product Design Specification Document**.

The Product Design Specification Document

This document contains all the facts related to the product. It provides qualitative information about the functional goals of the product and quantitative information defining product performance. The **Product Design Specification Document** is a statement of what the product has to do and is the fundamental control mechanism and basic reference source for the entire product development activity. This document forms the basis of specifications of the wheelchair as designed and manufactured. It should be submitted to the **User Panel** for suggestions and criticisms.

The **Product Design Specification Document** should contain:

- The product title.
- A general description including the product concept and strategic goals specifying why there is a need for the new product.
- A user profile and summary of product user needs.
- Design objectives for the product.
- An ergonomic analysis including a description of product function and dysfunction and a task analysis.
- The specification of the product user requirements and the corresponding engineering characteristics in the form of a QFD matrix.
- Design constraints related to cost, technology, regulations and standards, user capabilities, and the environment.
- Marketing requirements, including an analysis of what types of products it will be competing with, who makes them, and what market it will serve.
- The anticipated demand and target price.

Most of the components of the **Product Design Specification Document** stated above were described in the previous sub-sections, with the exception of some aspects such as strategic product goals, costs, and technology, which are not central to the remit of this book.

7.2.5 DESIGN CREATION

As product users' requirements become defined, there is a need to study the alternatives to satisfying these requirements in terms of the three-dimensional shape of the product. This phase of the ergodesign methodology involves generating solutions to meet the statements included in the **Product Design Specification Document**. The solution will represent the sum of all of the subsystems and their components that go to make up the whole system working as required to satisfy user needs. So, the first phase of the **Design Creation** process will start with a set of user needs and product specifications and will result in a set of product concepts/alternatives from which the design team and the **User Panel** will make a final selection.

The **Design Creation** phase is very complex in as much as it has several goals, many constraints, and an even greater number of possible solutions. The major challenge to the design team will be to design a new product in order to meet the needs of a wide range of users, exploiting to the full the abilities of sales, marketing, and distribution channels, fitting in with existing manufacturing facilities and suppliers and ending up making a profit for the company. Only professionals with design competence will be able to complete this phase.

ATTENTION

The **Design Creation** phase must be carried out systematically. It will be divided into the following steps:

- Generating concepts.
- Evaluating and selecting concepts.
- Refining concepts.
- Detailing design.

7.2.5.1 Generating Concepts

The **Generation of Concepts** phase is also known as Generation of Design Alternatives. It is essential to begin with the **Generation of Concepts** having the design problem sufficiently clarified. Clarifying the problem consists of developing a general understanding and then breaking the problem down into subproblems. This was previously done in the phase named **Investigating the Problem**. The **Analysis of Dysfunction of the Product–User Interface** revealed several problems that should be improved by the designers in the design of the new product.

The generation of new ideas is at the heart of the **Generating Concepts** phase. There is a wide range of techniques for the generation of creative ideas such as brainstorming and mind mapping, synectics, removing mental blocks, bionics and biomimetics, morphological charts, parametric analysis, and problem abstraction. The choice of which technique or techniques to use is a personal choice of the designers and will depend on which one they are more familiar with.

Each technique has its advantages and disadvantages. They are exhaustively described by Ulrich and Eppinger (2019), Baxter (1995), Jones (1992), and Rozenburg and Eekels (1995). A summary of some techniques for the generation of creative ideas is shown in Table 7.9. The designer also can use the support of the **User Panel** in this phase of the methodology.

Other methodologies are also used by designers: human-centered design and design thinking. **Human-Centered** or **User-Centered Design** are an approach to interactive product development that aims to make products usable and useful by focusing on the users, their needs and requirements, and by applying human factors/ergonomics, usability knowledge, and techniques. This approach enhances effectiveness and efficiency, improves human well-being, user satisfaction, accessibility, and sustainability, and

TABLE 7.9

Summary of Some Techniques for Idea Generation (from Forcelini, 2018; Baxter, 1995; Jones, 1992)

Technique	Procedures
Bionic and Biomimetic	• Look for references in nature for the functional solution of the problems. • Analyze the problem and try to find similar functions.
Brainstorming	• Select a group of people to produce ideas. • Enforce the rule that no idea is to be criticized and make it clear that wild ideas are welcome, quantity is wanted, and that participants should try to combine, or to improve upon, the ideas suggested by others. • Record the ideas put forward and evaluate them afterward. • The ideas can be represented as mind-maps which is a graphical technique for imagining connections between various pieces of information or ideas.
Brainwriting	• Select a group of people to write a limited number of ideas on a single sheet of paper, either in columns or rows. • Each sheet is then handed to someone else in the group and they have to try to improve or develop all of the ideas a step further by adding a new row or column until ideas have been exhausted or until each sheet has been round every group member. • Carry out a conventional session of brainstorming to bring out any completely new ideas not on any of the sheets but stimulated during the brainwriting process.
Mind Mapping	• Define the concept (words or signs that stimulate mental images), the link words (verbs), and the preposition (they serve to unite and relate concepts – *the, after, until, against, from, before, about*). • Graph the problem you are analyzing.
Morphological Chart	• Specify the problem • Select the problem parameters • Make a list of variations • Try different combinations.
Persona	• Describe the target audience, as realistically as possible, by creating imaginary people. • Describe the user's profile and analyze how and where to use the product. • From there, understand how and in what context the user will use the product to serve as a subsidy to product design. • Illustrate your persona graphically to represent the user's characteristics.
Removing Mental Blocks	• Transformation rules or design reorganization that can be applied to transform existing unsatisfactory solution or parts of it (e.g., put to other users? adapt? modify? substitute? reverse?) • Searching for new relationships between parts of an existing unsatisfactory solution. • Re-assessment of the design situation.
Synectics	• Form a group of highly selected people to operate as an independent development department. • Give the group a lot of practice in the use of direct, personal, symbolic, and fantasy analogies to relate the spontaneous activity of the brain and nervous system to the problem. • Submit to the group difficult problems that the parent organization cannot solve and allow plenty of time for solving. • Submit the group's output for evaluation and implementation.

(*Continued*)

TABLE 7.9 (CONTINUED)
Summary of Some Techniques for Idea Generation (from Forcelini, 2018; Baxter, 1995; Jones, 1992)

Technique	Procedures
Parametric Analysis	• Pick up an existing product that comes closest to solving the problem with particular attention being paid to the parameters in which the product fails to provide a complete solution.
	• Analyse the product features in terms of quantitative parameters (size, power, speed, strength, price, efficiency, durability), qualitative parameters (ranked or scaled to other products), and categorical parameters (categories the product belongs to).
	• Indicate how these parameters would have to be different to fully solve the problem.
Problem Abstraction	• Make a statement of the problem.
	• Ask "why" the design team wants to solve the problem.
	• The answer is then challenged with further "why" questions until the company's ultimate objective is reached.
	• Each level of abstraction should reveal a new set of potential solutions.

counteracts possible adverse effects of use on human health, safety, and performance. In this book, we expand this concept to **Human-Centered Design** that contemplates aspects related to human psychology and perception including the user's emotions, as explained in the preface.

NOTE

Want to learn more about User-Centered Design? See the following references:

Glacomin, J. (2014). What Is Human Centred Design? *The Design Journal*. V. 17 (4), doi.org/10.2752/175630614X14056185480186.
IDEO (2015). *The Field Guide to Human-Centered Design*. IDEO.org / Design Kit. ISBN-13: 978-0991406319.
ISO 9241-210:2019 (2019). *Ergonomics of human-system interaction – Part 210: Human-centred design for interactive systems*. Available at: https://www.iso.org/standard/77520.html. Accessed on 17 October 2019.
LUMA Institute (2012). *Innovating for People Handbook of Human-Centered Design Methods*. LUMA Institute. ISBN-13: 978-0985750909.
Still, B. and Crane, K. (2016). *Fundamentals of User-Centered Design: A Practical Approach*. CRC Press. ISBN-13: 978-1498764360.

Design thinking is a process for creative problem solving based on the human-centered core. According to Dam and Siang (2019), **Design Thinking** is an iterative process in which it is seeking to understand the user, challenge assumptions, and redefine problems in an attempt to identify alternative strategies and solutions that might not be instantly apparent with our initial level of understanding. The authors point out the importance of Design Thinking because designers' work processes can help systematically extract, teach, learn, and apply human-centered techniques to solve problems in a creative and innovative way – in the designs, in the businesses, in the countries, in the lives.

NOTE

You can find more information about Design Thinking at:

Curedale, R. (2019). *Design Thinking: Process & Methods*. 5th ed. Design Community College. ISBN-13: 978-1940805450

Lewrick, M.; Link, P.; Leifer, L. (2018). *The Design Thinking Playbook: Mindful Digital Transformation of Teams, Products, Services, Businesses and Ecosystems*. Wiley. ISBN-13: 978-1119467472

Lockwood, T.; Papke, E. (2017). *Innovation by Design: How Any Organization Can Leverage Design Thinking to Produce Change, Drive New Ideas, and Deliver Meaningful Solutions*. Weiser. ISBN-13: 978-1632651167.

Pressman, A. (2018). *Design Thinking: A Guide to Creative Problem Solving for Everyone*. Routledge. ISBN-13: 978-1138673472.

The objective of the **Generating Concepts** phase is to accumulate as many ideas as possible, so attempts to filter them at this stage should be suppressed. As idea generation comes from imagination and creativity, rational associations, commonly used in everyday life, should be avoided. Also, ideas that initially may appear not feasible can often be improved by other members of the design team. Designers should invite the **User Panel** to take part in some creative sessions to help in finding solutions to specific problems. The use of renderings and mock-ups to express the designers' ideas will be more appropriate than text and spoken language. Computer-aided industrial design (CAID) tools may also be used to generate three-dimensional designs on a computer screen, with the possibility of producing a great number of detailed concepts that can be rapidly modified. Figure 7.5 illustrates the phase of **Generating concepts** with the sketch of some solutions for the design of wheelchairs (Design bei Rollstühlen, 1993).

It is important to draw attention to the fact that, as opposed to the engineers' team who focus their attention upon finding solutions to the technical subfunctions of the product, the design team will concentrate upon creating the product's form and functionality and user interface. Naturally, the concepts resulting from the design team intervention should meet user needs and product specifications previously defined.

7.2.5.2 Evaluating and Selecting Concepts

Design is a process of divergence and convergence. As an evolutionary process, the design of a product grows from a product idea via solution principles, concepts, and preliminary designs to detailed and definitive design. A number of concepts are generated in each phase and need to be evaluated and selected in order to find the best solution to the design problem.

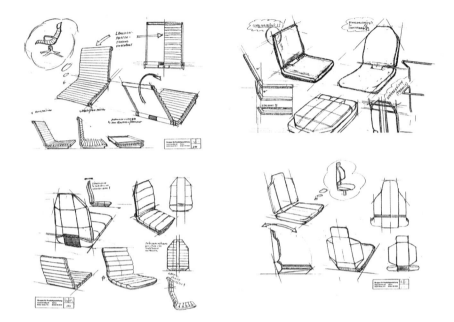

FIGURE 7.5 Samples of some sketches made to backrest design of a wheelchair (Design bei Rollstühlen, 1993).

ATTENTION

The purpose of the **Concept Evaluation and Selection** phase is to allow:

- The various design concepts/alternatives can be compared,
- Some alternatives are eliminated,
- Iterations can be performed so that new alternatives can emerge from the combination of some previously generated concepts, and
- Enable a new, more refined selection to choose the final concept.

In general, this stage of product design is performed subjectively. Thus, the objective of this phase of the ergodesign methodology will be to establish criteria in which a large number of functional and conceptual ideas will be filtered and selected in a form to choose the best options to meet user requirements and product specifications.

The use of matrices as a means of structuring or representing evaluation and selection procedures is advocated by a number of authors, including Ulrich and Eppinger (2019), Magrab (2009), Baxter (1995), Fox (1993), and Pugh (1991). In the **Ergodesign Methodology for Product Design**, it is recommended to use some or all of the user requirements (needs) identified in the previous phases as criteria to evaluate the design concepts.

TABLE 7.10

The Matrix for Evaluating and Selecting Concepts

Subsystem: Pushandle		Concepts					
Selection criteria	A	B	C	D	E	F	G
Ease of handling		+	0	0	0	0	0
Ease of use	R	+	0	0	+	0	0
Ease of removal	E	0	0	0	0	+	0
Maneuverability	F	0	-	-	0	+	-
Sharp edges are smoothed off	E	0	0	0	0	0	0
Reduction of vibration in the hands	R	-	0	0	0	+	-
	E						
Good stability	N	0	-	0	+	+	0
Adjustability	C	+	0	+	+	-	0
Help in curb climbing	E	0	0	0	+	0	0
Easy to fit accessories		+	0	0	-	-	0
Wheelchair foldable	C	0	0	+	-	0	0
Safety	O	0	-	0	0	0	0
Low manufacturing costs	N	0	0	0	-	0	+
Sum +'s	C	4	0	2	4	4	1
Sum 0's	E	8	10	10	6	7	10
Sum −'s	P	1	3	1	2	2	2
Net Score	T	3	−3	1	2	2	−1
Rank		1	5	3	2	2	4
Continue?		Yes	No	Yes	Combine	Combine	No

The **Matrix for evaluating and selection concepts** (Table 7.10, based on Ulrich and Eppinger, 2019; Magrab, 2009; and Pugh, 1991) works as a means to narrow and improve a number of product concepts.

This matrix should be used to analyze different aspects of the product, such as its subsystems, sub-subsystems, components, or combinations of them. Also, it would be extremely useful to analyze the esthetic aspects of the product and its components. It is indispensable the participation of the **User Panel** to help the design team to analyze certain aspects of the matrix, for example, aspects concerning esthetics and usability.

The steps involved in preparing the matrix, as in Table 7.10, include the following phases:

(a) Preparation

- Define the team which will take part in the evaluation and selection sessions, including the **User Panel**. Be sure that all members are supplied with enough information about the concepts to be evaluated and selected and a list of criteria to be used.

- Provide a computer with projector for figures, whiteboard, and or a flipchart if required.
- Provide appropriate support materials, e.g., pen and notepads.
- Be sure that each concept is presented in the form of sketches, renderings, mock-ups, and/or real or digital models, and they are all illustrated/presented with the same level of detail.
- Establish the **Selection Criteria** against which the concepts are to be evaluated and list them down the first column on the left-hand side of the matrix. The criteria should be based on user needs and the needs of the company, such as low manufacturing costs or product safety. Be sure that the criteria chosen are absolutely important, relevant, unambiguous, understood, and accepted by all participants in the evaluation and selection sessions.
- Choose a concept to become a **Reference Concept** against which all other concepts are rated. The reference may be: (a) a product, system or subsystem, which is commercially available and is agreed upon by the judging group as being of excellent quality; (b) an obvious solution to the problem; (c) an industry standard; and (d) in the case where competitive designs/concepts do not yet exist, any one of the concepts under consideration that the group agrees intuitively as the best choice. The **Reference Concept** is placed in column A of the matrix.

(b) Rating the Concepts

- Make a comparison between each concept and the **Reference Concept.**
- According to a team consensus, for each concept attribute **Relative scores**, such as "better than" (+), "same as" (0), or "worse than" (-) the chosen reference and in relation to each one of the selection criteria.
- Write down the relative score in each cell of the matrix which makes the intersection between the concept and the selection criteria on which is currently being analyzed, for instance, when the concept "B" was rated against the selection criterion "Ease of handling", it got the score "+".

(c) Ranking the Concepts

- Add the +'s scores ("better than") of each concept and enter the result in the appropriate cells in the lower row of the matrix.
- Do the same with the 0's ("same as") and −'s ("worse than") and enter the result in the appropriate cells.
- Calculate the **Net Score** subtracting the number of scores that have received "worse than" ratings from those which have received the "better than" ratings. Ignore those which have been rated "same as".

- **Rank Ordinally** the concepts which have received more pluses and fewer minuses. Those that received the highest ratings are the ones that should continue to the next stage of **Refining Concepts**.

(d) Combining and Improving the Concepts

- Observe if there are some concepts that although they are good in some way, are still not suitable as the best design solution, because they have some bad features.
- If affirmative, consider if these concepts may be combined. E.g., in the sample in Table 7.10, concepts E and F can be combined to form a new concept (Concept EF) and will be considered in the next phase of **Refining Concepts**.

(e) Select the Concepts and Reflect on the Results

- Decide with the other participants (including the **User Panel**) which concepts are to be selected for further refinement.
- Reflect on the result of the process.

As previously mentioned, the design team will use mock-ups and/or real and digital models to represent their design concepts. The terms mock-up, model, and prototype may sometimes be ambiguous. In this book, the word "model" refers to a functional and physical representation of the entire product as it will eventually be manufactured.

"Models" represent the size and shape of a subsystem or component of a product. They are usually produced on a much smaller scale than the actual product but with no necessary relation to function and/or appearance. Mock-ups, on the other hand, represent the size, shape, and appearance of a product, subsystem, or component, normally executed with different material and cheaper than the original and made in full scale, but unrelated to the function.

The "prototypes" are functional representations of the product presented on a full-scale scale. They differ from models and mock-ups as they are usually built in a smaller size and according to the level of detail necessary to represent the types of planned evaluations and static simulations. The final prototype is considered as the first physical object that represents the final product, with all its features and functionality.

Mock-ups and/or real or digital test models are used to assess the feasibility of specific design concepts, with the aim of validating the concept and identifying any obvious or predictable problems before incurring the cost of building a functional prototype. Please note that in product analysis only full-size models are useful for usability evaluation. Thus, mock-ups and models, unlike prototypes, can be made of paper, foam, wood, or any other material without a necessary connection to what will be used in the final version of the product. Figure 7.6 presents examples of some wheelchair models designed to represent different design concepts (Design bei Rollstühlen, 1993).

FIGURE 7.6 Examples of some models of wheelchairs representing different design concepts (Design bei Rollstühlen, 1993).

7.2.5.3 Refining Concepts

Refining concepts are used to help in the final decision to select one or more concepts able to be developed. A **Matrix of Refining Concepts** (Table 7.11), similar to the previous **Matrix for evaluating and selecting concepts**, is built using the following steps:

(a) Preparation

- Similar to the previous matrix, each concept is presented in the form of sketches, rendering, mock-ups, and/or real or digital models and may include more details to express its forms and functions
- Establish the **Selection Criteria** against which the concepts are to be evaluated as was previously done. Most of the criteria should be the same as those used in the previous **Matrix for evaluating and selecting concepts** and, if appropriate, can be deployed to help the assessment.
- With the help of the **User Panel** and under using a consensual approach, attribute subjective weights to each criterion in the form of percentages. The weights are listed in a column after the column containing the **Selection criteria**. The subjective weights are so constructed as to sum to 100%.
- Choose a concept to become a **Reference** against which all other concepts are rated in a similar way as in the **Matrix of evaluation and selection concepts** (Table 7.10).

TABLE 7.11
The Matrix of Refining Concepts

Subsystem: Pushandle

			Concepts					
		A	B		D		EF	
Selection criteria	Weight		Rating	Score	Rating	Score	Rating	Score
Ease of handling	5	R	3	15	2	10	4	20
Ease of use	5	E	2	10	2	10	3	15
Ease of removal	5	F	1	5	1	5	3	15
Maneuverability	10	E	3	30	2	20	4	40
Sharp edges are smoothed off	5	R	3	15	3	15	3	15
Reduction of vibration in the hands	5	N	3	15	3	15	3	15
		C						
Good stability	10	E	4	40	3	30	4	40
Adjustability	10		3	30	2	20	5	50
Help in curb climbing	5	C	4	20	1	5	4	20
Easy to fit accessories	5	O	3	15	3	15	4	20
Wheelchair foldable	5	N	3	15	3	15	4	20
Safety	15	C	3	45	3	45	5	75
Low manufacturing costs	15	E	1	15	3	45	5	75
Total Score		P		270		250		420
Rank		T		2		3		1
Continue?				No		No		Develop

(b) Rating the Concepts

- Make a comparison between each concept and the chosen **Reference Concept** for each selection criterion
- According to a team consensus, for each comparison must produce a classification, presented in the column "Concepts", for each product evaluated. The concepts can be:
  1 = much worse than reference concept;
  2 = worse than reference concept;
  3 = same as reference concept;
  4 = better than reference concept; and
  5 = much better than reference concept.
      For example, when the concept "B" was rated against the **Reference Concept** for the selection criterion "Ease of handling" got the rating of 3.
- For each selection criteria and each concept, multiply the "weight" by the "rating" and write down the value in the **Score** column. For example, completing the case described under the previous bullet point, the rating of 3 is multiplied by a weight of 5 to give a score of 15.

(c) Rank the Concepts

- Add the scores for each concept, e.g., the total score for concept "B" is 270, for "D" is 250, and for "EF" is 420.
- Order the concepts, with that obtaining the highest total score in the first place and the lowest total score in the last place.

(d) Combining and Improving the Concepts

- As in the previous matrix, observe if there are ways to combine and improve good concepts.

(e) Select the Concept and Reflect on the Results

- Decide, in consensus with the participants, which concept is to be selected for further development; and
- Reflect on the result of the process and be sure that the concept is in accordance with what was previously established in the QFD matrix.

7.2.5.4   Detailing Design

The objective of this phase of the **Ergodesign Methodology for Product Design** is to show, in the form of technical drawings (using either paper or computer software), that the chosen concept has its properties detailed sufficiently well to be modeled and/or prototyped and manufactured.

Products comprise certain properties, however, only some of the product properties can directly be determined by industrial designers. Examples of some properties which can be determined by industrial designers include the structure of the entire product (the arrangement of the parts), the shape, the dimensions, the material, the colors, the surface quality, and texture. Properties that are not usually part of an industrial designer's body of knowledge include the analysis of tolerance, corrosion resistance, strength and durability of materials, the choice of the manufacturing method, the analysis of product value. Most of the product properties, in terms of specifications, have been defined since the first phases of the design process and have already been incorporated in the concept choice.

This phase of the design process includes progressive levels of complexity. This complexity varies between the production of sketches, mock-ups, and/or real or digital models made in the conceptual phase and the more detailed specification of materials, principles, and manufacturing process required in the design of a final prototype (the master copy of the product which will later be mass-produced). Part of the **Detailing design** phase can be considered as included in the product development and production planning of the product which is mentioned briefly in the next step of this design methodology. These steps involve the participation of other technical professionals, such as manufacturing and mechanical engineers.

Based on sketches and models built in the previous design phase, the industrial designer will draw and detail the product's geometrical shape, dimensions, material, color, arrangement of subsystems, and components. Details related to some aspects of the product, such as esthetic appeal, safety, user interface, and product maintenance, should be carefully specified. Intermediate stages in the **Detailing design** may require that designers produce models to check the accuracy of data. The **Detailing design** phase also involves a decision about which components will be bought in (i.e., selection and purchase of standard catalog items) and which will be manufactured, either in-house or by sub-contractors.

It is important to mention that in this phase of the design process, all the subsystems and components, which represent user needs, functionality, and style, should be brought together and integrated into the whole product ready to be manufactured.

Two other design activities are also included in this phase of the **Ergodesign Methodology for Product Design**: the **Design of the user manual** and the **Design of promotional material**.

7.2.5.5   The Design of the User Manual

This step of the methodology aims to prepare the **User Manual**, together with the marketing and the graphic design team, in order to provide instructions on the operation of the product. The prevalence of inadequate user manuals contributes to the fact that users may ignore important information or simply avoid consulting the manual. A clear understanding of user needs and the way they perform using the product is the first step to enabling the design team to develop an adequate user manual. **Task Analysis** is an excellent tool to assist in creating the **User Manual**.

TIP

Recommendations for the design and writing of user manuals can be found in several authors, including Hodgson (2019), Pavel and Zitkus (2017), Cifter (2010), Cushman and Rosenberg (1991), and Weiss (1991).

Møller (2013) stated that applying user tests as a supplement to guidelines and proofreading when developing user manuals has many advantages. Testing of the manual by a sample of users will ensure that understanding the use of the product will prevent any errors and frustrations in its use.

Wiese et al. (2004) studied the consumers' use of written product information and found that self-report data, collected on a wide range of products, suggest that product complexity is the best predictor of instruction manual use. In a text to the "BBC Future", Schumacher (2019) states that new technologies such as Artificial Intelligence (AI) and Augmented Reality (AR) are also starting to be used to instruct users to use products and AR can allow instructions to be layered so users can interact with a product while learning how to use it. The author also states that AR can allow instructions to be described in layers so that users can interact with a product while learning to use it.

Acioly (2016), in a doctorate thesis which I supervised, studied the application of Mobile Augmented Reality technology as a way of orientating the use and security instructions (alerts and warnings) in the users of consumer sardine can packaging. The study was focused on evaluating the measurements of usability (efficiency, effectiveness, and satisfaction) of two information systems, one physical (label) and the other digital (Augmented Reality application). The results show whereby the presentation of the instructions of use and security through digital information systems in Moblie Augmented Reality on consumer packaging presents itself as an efficient and effective solution as an interface of communication and interaction with its users.

In a study carried out by Li et al. (2018) on Chinese Technical Communicators' Opinions on Cultural Differences Between Chinese and Western User Manuals, it was found that that the majority of the interviewees assumed that Chinese and Western manuals differ from each other in many aspects (content, structure, style, visuals) and that Chinese and Western users have a different approach to user manuals. This draws attention to the importance of considering cultural aspects in user manual design for products that are marketed in several countries.

Cushman and Rosenberg (1991) recommend the following steps for the design of good user manuals:

- Organize material in a logical manner consistent with reader expectations.
- Provide adequate structure (e.g. different type styles and sizes for main headings and subheadings, use of spacing for demarcation, descriptions in margins, highlighting, etc.).
- Present only information that the reader will need.

- Use words that the reader will understand.
- Use simple sentences and active voice.
- Present sequential instructions and procedures in lists, outlines with "bullets", or flow diagrams rather than in paragraph form.
- Use figures to help clarify the message.
- Place figures and accompanying verbal explanations on the same page or facing pages.
- Test, revise, and retest the user manual until novice users can perform all tasks without difficulty.

As a product which may have among its users a significant number of people with poor cognitive ability, designers should pay special attention to the design of user manuals for disabled users, including the use of large and sans-serif fonts, the provision of illustration wherever feasible, and the provision of text description for all illustrations. The design team should submit a draft of the user manual of the new product for the appreciation and criticism of the **User Panel**.

7.2.5.6   The Design of Promotional Material

The **Design of promotional material** is part of the company's marketing strategy and helps to promote its image, thus contributing to increase and retain the loyalty of the target audience. The designer, specialized in graphic design, and the marketing team must work together to create strategies for promoting the new product

7.2.6   PROTOTYPING

Although a number of design and engineering problems can be solved using computer simulation, drawings, mock-ups, and models, building a physical real size prototype as a functional representation of the final product permits the design team to test and evaluate the design concept and usability aspects of the product. The real size prototype will help to evaluate the product performance in terms of meeting the required specifications and user needs and will reveal problems that arise from the engineering of the product. The final prototype is built with the same materials as the final product and it will be the first product produced in the series of products that will follow.

It is expected that tests with the prototype will identify any remaining problems in terms of product specification that was not be identified in the previous phases of design. Otherwise, a large amount of money and time may be spent later in the production process to remedy any failures. An example of a prototype of a wheelchair built to test product performance is shown in Figure 7.7 (Design bei Rollstühlen, 1993).

The building of a prototype for the new product concept will be useful for:

- Learning if the concept represented by the prototype will work and meet the user needs and the product specifications;
- Communicating the product concept and its features to the top management personnel, partners, vendors, users, and other members involved in the

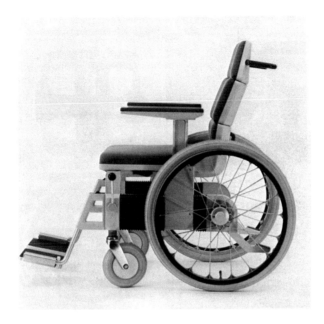

FIGURE 7.7 Example of a prototype of a wheelchair built to test product performance (Design bei Rollstühlen, 1993).

product development process. It is easier to obtain feedback on the product using a visual, tactile, and three-dimensional representation than by verbal description or even sketches and drawings of the product;

- Integrating the subsystems and components of the product in such way as to ensure that they work together as expected;
- Testing and verification of the new product in terms of the user–product interface and assembly and interconnection of all parts;
- - check if the usability aspects have been met and if the product provides a good user experience.
- Checking if safety and legal issues are satisfied;
- Assuring that raw materials and purchased components will meet performance and delivery requirements; and
- Checking if costs and production schedules will be within specified limits.

7.2.7 TESTING AND VERIFICATION

The product **Testing and Verification** step is usually carried out throughout the phase of **Design Creation**, starting with an evaluation of the first mock-ups and of the design and engineering models, and it concludes with verification tests of prototypes at field sites.

Indeed, one of the major aims in building a three-dimensional representation of the product is to compare objective user-performance data obtained from the test with the product specification. In such a way, **Testing and verification** is a critical phase to improve product usability and quality, to reduce the likelihood of legal action against the product's manufacturer, and to contribute to the success of the product in the marketplace.

In this methodology, the word **Testing** is used to refer to those procedures which take place in a laboratory or other controlled environment. **Verification** refers to those tests that are carried out in a field environment rather than in a laboratory. A review of the Modeling and Prototyping phase is presented in sub-chapter 3.4, of this book.

ATTENTION

The **Testing and verification** step should consist of the following phases:

- Evaluation of the prototype through usability tests.
- Modification and re-testing of the prototype to correct any flaws identified during usability tests.
- Preparation of the technical report with the final result of the tests with the product prototype.

Although physical tests are essential to verify the product's technical quality, such as fatigue tests, they are not part of the objective of this book. The **Ergodesign Methodology for Product Design** will focus on usability tests involving representative product users (the **User Panel**) and working prototypes. Figure 7.8 illustrates an example of testing the technical, functional, and handling properties of a wheelchair prototype.

Hekstra (1993) introduces recommendations for wheelchair testing. They are:

- The user-wheelchair interface with respect to dimensions and operations;
- The performance of the wheelchair with respect to matters such as rolling resistance and maneuverability;
- The performance of the wheelchair with respect to safety, including its stability and the efficiency of its brakes; and
- The technical quality of the wheelchair under different conditions of use involving strength and durability requirements.

It can be considered that this test program presented by the author can also be adapted to be applied to other products, in addition to wheelchairs.

ATTENTION

In the **Ergodesign Methodology for Product Design**, the evaluation of physical prototypes includes usability tests and analysis of the user experience. Both must be conducted in this **Testing and Verification** step.

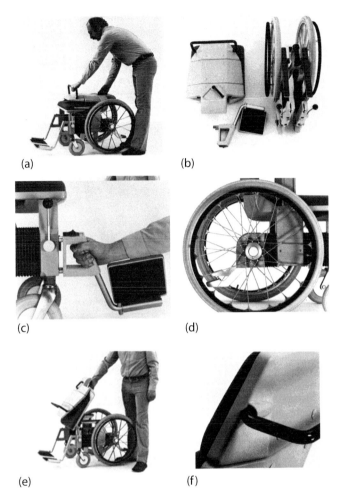

(a)   (b)

(c)   (d)

(e)   (f)

FIGURE 7.8 Example of testing of functional, handling, and technical properties of a wheelchair prototype (Design bei Rollstühlen, 1993).

The steps to perform the **Physical prototype evaluation** phase (usability tests and product prototype verification) are presented below, based on Soares and Rebelo (2017); Karwowski et al. (2011a, b); Dumas and Redish (1999) and Cushman and Rosenberg (1991).

7.2.7.1 Planning the Usability Test and Prototype Verification

- Provide the facilities in which the tests will be carried out. The facilities here are similar to those used in the focus group sessions, and the same recommendations for the selection and choice of facilities should be applied, as previously presented.

- Define the resources (people, equipment, time, financial resources, etc.) that should be devoted to the **Testing and verification** phase. Members of the **User Panel** should be invited to take part in the tests.
- Carry out a literature review on standards and previous tests of this or other similar products.
- Establish the aims of the **Testing and verification** including what will be measured and the criteria adopted for the adoption of: a) objective measures (objective measures, e.g., time to complete a task and error rates); and b) subjective measurements, e.g., user's perceptions, opinions, and judgments).
- Select the tasks that users will perform, including assembly, storage, maintenance, and following instructions in the user manua, bearing in mind that the tests will probe areas of potential usability problems. Information obtained from task analysis, interviews, and focus groups can help the design team to set what to measure.
- Establish qualitative (subjective measurements) and quantitative criteria (objective measurements) for measuring performance, focusing on users and not on the product.
- Define the duration of sessions and tests, taking into consideration the product complexity, the objectivity of the tests, the number of participants involved, and the length of time each task will be performed by the user.
- Decide on the test scenario, which means the description of tasks to be carried out in a way that takes some of the artificiality out of the tests. The scenario will tell the participants what they will do during the test.
- Decide where the tests will be performed either (a) in a laboratory or other controlled environment (Test), and/or (b) in a field setting (e.g., users' home or public premises – Verification). Select the methods and tools used in usability and user experience evaluation (a list of methods and tools is available in sub-chapters 6.4 and 6.5).
- Define the techniques used for observing and recording the tests, including video recording, automated data collection, questionnaires, and focus group sessions.
- Organize files with name and data of each participant, to register their performance doing the tasks.
- Be sure that the **Concept of Minimal Risk** is strictly observed. **Minimal Risk** means that "the probability and magnitude of harm or discomfort anticipated in the test are not greater, in and of themselves than those ordinarily encountered in daily life or during the performance of routine physical or psychological examinations or tests" (Dumas and Redish, 1999, p. 205).
- Obtain **Written informed consent** from the participants stating that they are aware of the procedures the tests will follow, the purpose of the test, any risks involved, the opportunity to ask questions, and the opportunity to withdraw at any time.

The use of the findings from the **Analysis of Dysfunction of the Product–User-Interface** in the phase of **Delimiting the problem** should be extremely useful to investigate whether the problems previously encountered have been overcome.

7.2.7.2   Conducting the Usability Test and Prototype Verification

- Greet the participants and create a relaxing atmosphere calming any fears or anxieties the users may have in testing the new product.
- Explain the test scenario to participants being sure that they clearly understand all the tasks to be performed.
- Take special care if the tests will be performed in the user's home or public premises, considering that other persons may be present at the location, and the participant may be shamed or embarrassed.
- Ask participants to think out loud so the design team can hear and record their reactions to the product. Give users instructions on how to think out loud as if they are alone in a room and one or two warm-up exercises before they start to perform the required task.
- Use checklists with the task scenario.
- Register, on an appropriate form, the results of the tasks as they are being performed.
- Register, on an appropriate form, any unforeseen problems that may appear.
- Record the whole session using a video camera.

7.2.7.3   Analyzing the Results

- Tabulate the data and use, if appropriate, statistics to describe the findings of the data.
- Summarise the findings organizing the problems (a) by subsystems, (b) prioritizing those who have the widest scope, and (c) organizing them by the level of severity.
- Analyse and interpret the results making clear if the prototype is meeting or not user needs.
- Propose recommendations.

7.2.7.4   Reporting the Usability Test

- Organize a meeting with the managers and product development team, including the User Panel, to show the results of the tests and prototype verification.
- To support the presentation, the design team may use figures, graphics, and a highlighted video recorder, including the most important findings.
- In addition to the verbal presentation, write a **Technical report** to the company's managers and other members of the product development team with the findings of the **Testing and verification** phase in a similar way to that made to report the focus group findings. This written report will

constitute the documentation of this phase of the **Ergodesign Methodology for Product Design**.

7.2.8 THE PHASES OF PRODUCT PRODUCTION AND MARKETING

Product design is ready for production if all design properties have been specified definitively and with all required details. Although the manufacturing process should have been considered in the later phases of concept and detail design, and prototyping, the product must be entirely specified for the manufacturing process.

According to Magrab (2009), there are basically three very important and inextricably linked elements in the product development cycle: (1) assembly methods, (2) manufacturing process, and (3) material selection. These greatly affect the final product's cost, marketing time, plant production, degree of manufacturing automation, productibility, and reliability. The QFD technique should continue to be used throughout the product development and manufacturing processes to guarantee that the users' voices will continue to be heard (see sub-chapter 6.2).

The final phases of the **Ergodesign Methodology for Product Design** – product production (including manufacturing and assembly), product marketing, and customer support – are not directly involved with the design and, in view of this, are not discussed in this book.

7.3   INVESTIGATING THE SUITABILITY OF THE PROPOSED METHODOLOGY

A sample of four designers who had previously participated in the field study at the start of the work was approached. This relatively small number was due to the time available, which made it impossible to have more respondents involved. The aim was to collect their views on the extent to which the proposed methodology was acceptable to them. The companies the designers worked for were all located in the UK.

The details of this phase of consulting designers to confirm the suitability of the proposed methodology are described in Appendix 7.

A revised version of the flowchart of the **Ergodesign Methodology for Product Design** including the recommendations made by the designers is shown in Figure 7.9. Each step of the methodology is connected by a black arrow. Observe that, compared with the previous version (Figure 7.1), the methodology has undergone the following changes:

- A **Dealer Panel** was included to provide views on some phases of the design process including the **Preliminary Strategic Planning** and **Testing and Verification**. This action reinforced the concept of co-design adopted in this book.
- There is a link between the phases of **Testing and Verification** and **Design Creation,** which permits a loop between these two phases and the phase of **Prototyping**.
- There is a link between the phases of **Manufacturing and Assembly** and **Preliminary Strategic Planning** (this will allow checks to be made on

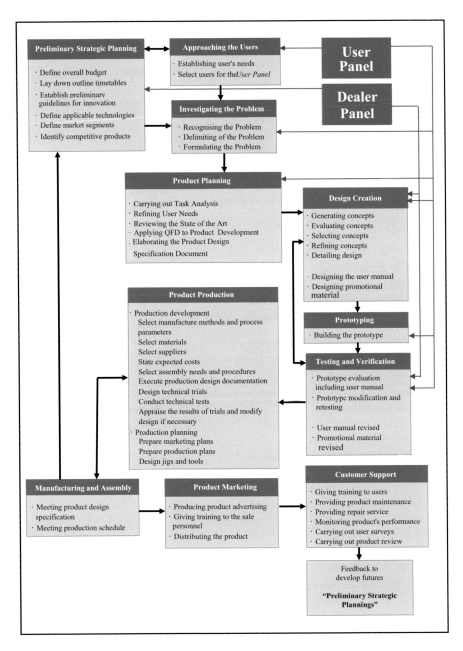

FIGURE 7.9   A revised version of the Ergodesign Methodology for Product Design. Source: The Author.

ously established).

- The phase of **Manufacturfeedbackembly** feedback to the **Product Production** phase (two-way arrow) to permit the checking of the first manufactured units against what was established in **Production Planning**.
- The feedback of the **Customer Support** phase will generate inputs to the development of future products.

We will present below a step-by-step summary for the use of this final version of the Ergodesign Methodology for Product Design.

7.4   STEP-BY-STEP SUMMARY OF THE ERGODESIGN METHODOLOGY FOR PRODUCT DESIGN

The following step-by-step summary is designed to facilitate the use of the methodology by professionals or students. A graphical representation will be presented below (Figure 7.10). The numbering of each stage and sub-stage below corresponds to the same presented in the graphical representation. This flowchart is linear, to understand the interactions between the various steps, see Figure 7.9.

1. **Preliminary Strategic Planning**

   At this stage, a series of decisions are made by the company's board of directors, such as the definition of the business plan, identification of the relationship between the new product and the company's other products, the definition of the costs of the product development program, project schedule, preliminary guidelines for product innovation, the definition of the technologies to be used, identification of the target market and competitive products and the definition of the starting point for the beginning of the product development process.

2. **Approaching the users and other stakeholders**

   This phase will be responsible for investigating existing information about the direct and indirect users of the product; develop user profiles, contact and select direct and indirect product users to participate in focus group sessions, conduct focus groups, select a sample of users to participate in the User Panel.

   2.1 Investigate existing information about product users

   Existing information about direct and indirect users within the company should be sought.

   2.2 Develop profiles of direct and indirect users.

   Objective: (a) define the profile of direct and indirect users; (b) contacting direct and indirect users; (c) define how direct and indirect users will participate in the methodology; (d) decide how many participants will be included and (e) recruit participants for focus group sessions.

   2.3 Conduct and analyze focus group sessions

   Using the techniques mentioned earlier in this chapter, conduct focus group sessions and define users' needs.

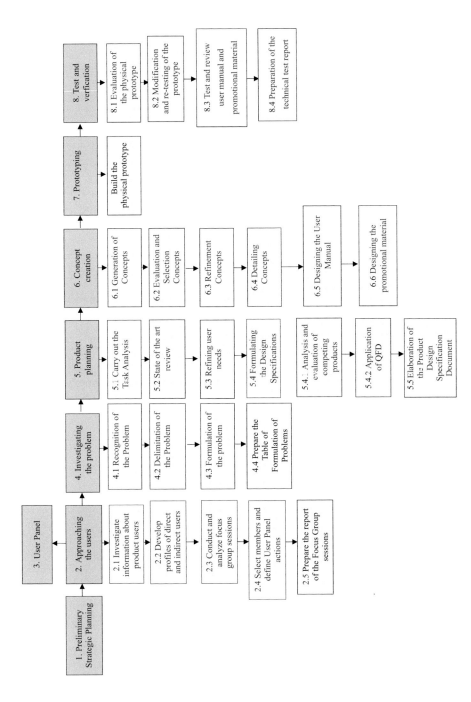

FIG. 7.10 Flowchart of the steps of the Ergodesign Methodology for Product Design. Source: The Author.

2.4 Select members and define User Panel actions

Select a sample of participants whose profile is representative of the population of end-users of the product in question and define their participation in the other stages of the methodology.

2.5 Prepare the report of the Focus Group sessions.

3. **User Panel**

The User Panel is a selected group of users who will participate, in a participatory way, in the various stages of this co-design methodology.

4. **Investigation of the problem**

This step involves identifying and categorizing any problems that are related to the use of the product. It consists of three phases:

4.1 Recognition of the Problem

It is intended to describe what is missing from the product or the situation to be analyzed or that exists but does not meet the needs of users. It corresponds to the investigation of the most serious and flagrant problems that appear in the first analysis of the problematic situation. It is suggested that it be carried out through focus groups with the User Panel. This step generates a Problem List obtained in a non-systematic way.

4.2 Delimitation of the Problem

At this stage, the previous recognized problems are selected, categorized, and classified. The problem categories are presented through the User–Product Interface Dysfunction Analysis. There are three types of dysfunctions in which problems can be categorized:

–   Ergonomic dysfunctions – Interface, postural, dimensional, instrumental, informational, perceptual, control, communicational, operational, cognitive, movement, natural, and accidental problems.

–   Human dysfunctions – Sensory-physiological (vision, hearing, touch) and psychoneurophysiological problems.

–   Machine malfunctions – Structural and moving, resistance and productivity, socio-cultural, semiological, and sensory-formal problems.

4.3 Formulation of the problem

The problems are reduced for those who have more significant and solvable aspects, considering the competence of the staff, the knowledge available, and what was required by users and the company. It is presented through a table called Formulation of Problems (4.4).

5. **Product planning**

This step aims to find information relevant to the subsequent activities of the new product design. For this, the Task Analysis, the State-of-the-Art Review, the Refining of User Needs are performed. The Product Planning step will generate the Product Design Specification Document.

5.1 Carry out the Task Analysis

–   It aims to identify and break down the activities that make up the task into as many observable activities as possible. It is presented through a flowchart called Task Activities Flowchart (Figure 7.3).

5.2 State-of-the-art review

Through a bibliographic search, we seek to find technical reports, books, magazines, periodicals, and conference proceedings with relevant articles that can contribute to the knowledge involved in the design of the new product.

5.3 Refining user needs

It aims to translate the user's needs previously identified in Product Requirements, in the Problem Formulation Table (Table 7.5), with clear specifications on "what" the product should do, how, in a measurable way to meet the user's needs. From the Problem Formulation table, problems must be: (a) associated with each subsystem of the new product, (b) assigned a specific importance level for each problem (from 1 to 5, where 1 is not very important and 5 is very important), c) presented the metric in which need will be measured and d) its respective unit of measurement. The result of this step will be presented in a table called List of Refined Needs of the user and its associated metrics (Table 7.6).

5.4 Formulating the Design Specifications

This phase comprises two stages presented below:

5.4.1 Analysis and evaluation of competing products

At this stage, products from competing companies are analyzed, evaluated, and compared with the company's own products in order to determine their strengths and weaknesses. This phase will generate two Competitive Product Tables: the first based on metrics and the second based on user satisfaction. These tables are based on the Refined Needs List (Table 7.6).

– Tables of Competitive Products based on Metrics (Table 7.7)

This table aims to compare the measurements of competing products for each metric identified. It must contain: (a) the number of each metric, (b) the number of the need, (c) each metric that was previously identified, (d) its level of importance, (e) the unit to be measured, and (f) the products of the competing companies that were analyzed and the measurements found for each metric.

– Table of Competitive Products Based on User Satisfaction (Table 7.8)

It is a table based on subjective evaluations of the level of satisfaction of the evaluators with each of the identified needs. Satisfaction evaluations are scored from 1 to 5. The table contains: (a) the need number, (b) the identification of each need, (c) the level of importance attributed to each need by users participating in the assessment, and (d) the assessment of the satisfaction of each need for each assessed company.

5.4.2 Application of the Quality Function Deployment (QFD) to Product Development

This phase aims at building the QFD Matrix (Figure 7.4). It consists of the construction of the 12 regions of the Casa da Qualidade, extracted from the previous tables, which include: (a) User requirements, (b) Classifications of importance, (c) Metrics or engineering characteristics, (d) Correlation matrix, (e) Matrix of relationship, (f) Competitive user evaluation, (g) Absolute importance, (h) Relative importance, (i) Measurement unit, (j) Competitive technical evaluation, (k) Target values, and (l) Technical difficulty.

5.5 Elaboration of the Product Design Specification Document

It is a document that must contain information related to the Strategic Planning of the Product, including the Analysis of the Task, Review of the state of the art, Definition of the needs of the users, Evaluation of the Competing Products, and the result of the Application of the QFD.

6. **Concept creation (Design)**

This stage is responsible for generating solutions and design alternatives to meet the Product Design Specification Document. This stage consists of four phases: (a) Concept generation, (b) Concept evaluation and selection, (c) Concept refinement, and (d) Concept detailing (Design).

6.1 Generation of Concepts

It consists of generating as many creative ideas as possible to solve the design problem. It is a phase that involves the creativity of the designer to solve the problems previously identified and meet the needs of users.

6.2 Evaluation and Selection Concepts

This step is responsible for establishing criteria to evaluate and select concepts that represent the best solution to the design problem. User Requirements are used as criteria for assessing design concepts. This is done through the Matrix for Concept Evaluation and Selection (Table 7.10). The matrix uses a Reference Concept with which the Selection Criteria are evaluated to define which design alternatives are considered better, equal, or worse than those of the Reference Concept. Each subsystem must be evaluated and classified. The concepts can be combined and improved through objective criteria and indicated for further improvement.

6.3 Refinement Concepts

Aims to help in the final decision to select one or more concepts that can be developed. A Concept Refinement Matrix (Table 7.11) is created similarly to the previous matrix. The concept chosen as a design solution must be in accordance with what was previously established in the QFD matrix.

6.4 Detailing Concepts

The chosen concept is presented in the form of technical drawings (using paper or specific computer software). The properties of the chosen concept must be sufficiently detailed to be modeled and/or prototyped and manufactured.

6.5 Designing the User Manual

This phase aims to prepare the User Manual, together with the marketing and graphic designer teams, in order to provide instructions on the operation of the product.

6.6 Designing the promotional material

It aims to create strategies, together with the marketing team, for the promotion of the new product.

7. **Prototyping**

In this stage, a physical prototype will be built, in real size, with a functional representation of the final product.

8. **Testing and verification**

This step, which can be performed throughout the Design Creation phase through test models and must include usability tests and user experience analysis. It consists of the phases:

8.1 Evaluation with models and mock-ups

Performed during the Design Creation phase to test the suitability of the proposed design alternatives. May use physical or digital models for testing.

8.2 Evaluation with the physical prototype

The evaluation of the physical prototype is done through usability tests and analysis of the user experience, which must follow previously established protocols.

8.3 Modification and re-testing of the prototype

After testing the usability and user experience tests, the recommended modifications and new tests must be carried out to meet the recommendations in order to overcome any problems identified.

8.4 Test and review user manual and promotional material

Conduct tests and review of the User Manual and Promotional Material together with the User Panel.

8.5 Preparation of the technical test report

A technical report must be prepared with the results of the tests and evaluations.

The other phases of product development are composed of Product Production, Manufacturing and Assembly, Product Marketing and Consumer Support and do not directly involve the participation of designers and ergonomists, because of this they are not part of the scope of this methodology.

The next chapter presents the conclusion of this book with reflections about the proposed methodology steps and their various phases of product development.

8 Conclusion

An overview of the book, its reflections, and proposals are presented. The chapter discusses the methodology developed and its phases, the safety of products developed for the disabled and non-disabled population, and the limitations of the study conducted by the author. It concludes that the use of this co-design human-centered methodology for product design is an effective option to guarantee that product direct and indirect users have their voice heard and their needs incorporated in a product that will have a good chance of providing full consumer satisfaction.

Design methodologies are procedures that provide designers and other professionals involved in product development with ways to guide the design according to specific approaches. Such approaches may prioritize consumer needs, specifications, or technological and manufacturing costs. This book presented modern design and manufacturing techniques using a human-centered approach and a co-design methodology. In this way, it aimed to solve design problems based on the needs of those who actually use the product.

Even when not considering aspects such as costs and manufacturing in-depth, the **Ergodesign Methodology for Product Design** has the virtue of allowing user needs to guide the steps of design so that a product is made which fully provide consumer satisfaction.

Industrial designers and ergonomists are directly involved in the six phases of the methodology, including **Approaching the Users**, **Investigating the Problem**, **Product Planning**, **Design Creation**, **Prototyping**, and **Testing and Verification**.

The methodology was introduced to four designers to see to what extent it was acceptable to them. Although the designers had just about an hour to evaluate, they unanimously responded in a positive way to validate the methodology. The criticism and suggestions that they made did not affect the essence of the methodology and were incorporated in a revised version shown in Figure 7.8. One useful suggestion was to incorporate a **Dealer Panel** in addition to the existing **User Panel** confirming the co-design characteristic of this methodology. As we suggested previously, this methodology can be used as an academic exercise. Obviously, in this case, the Dealer Panel is not applicable.

The need to produce a co-design methodology for product design that considers and involves users in the different phases of the process was considered positive by all the designers interviewed by the author. The designers interviewed were involved in the process of product design, evaluation, and use.

Products are designed to be used and to provide pleasure and satisfaction to their users. Additionally, disabled people need products that meet not only their medical and therapeutic needs but also improve their independence, quality of life, and give pleasure and satisfaction. Designing products for disabled people having in view just sales and profits, is at best short-sighted and worst immoral. Designing products

DOI: 10.1201/9781003214793-8

solely for the non-disabled population without considering those with physical and cognitive disabilities is a form of design discrimination.

For how long will products continue to harm and kill people? For how long will people continue to be unaware of the risk of exposing themselves and their families to the danger of potentially unsafe consumer products? For how long will millions of non-disabled users continue to have problems in using consumer products in general and disabled people wheelchairs in particular? For how long will designers continue to be insensitive to the voices of users (either non-disabled or disabled) and to translate their needs into product design? Unfortunately, this book is not able to answer these questions.

At first sight, it may sound utopian and not financially feasible to incorporate enough features in a product to achieve a design that satisfies a large range of users or to produce a design with enough modularity to incorporate the needs of many individuals. Certainly, there will continue to be limitations in such an approach so that people with more severe disabilities cannot be served. However, incorporating features and modularity will help countless disabled users who are sure to give a positive return on the initial financial investment made in design and manufacturing. Indeed, the creativity, imagination, and technical competence of designers will help in finding adequate solutions and new concepts able to balance the distinct requirements of user needs and manufacturing, marketing, and financial requirements. But an essential point for designers and manufacturers is to discard the preconceptions and false distinctions about who is young and old and who is non-disabled and disabled.

The design methodology provided in this book derived from the lessons learned from the literature review and the surveys of the stakeholders involved in the processes of design, prescription, and use of wheelchairs was an answer to the question "how to translate user needs into product design for general users, and wheelchair users in particular?".

The **Ergodesign Methodology for Product Design** has incorporated consolidated design practices found in the literature with the innovative incorporation of users throughout the several phases of the co-design process. As was said before, a design methodology itself is not sufficient to guarantee the good quality of the design of any particular product or the success of the product in the marketplace. But the risks as well the costs can be minimized by following good practice and incorporating the users' needs to achieve consumer satisfaction. The only way to guarantee that a design methodology represents good practice is if designers accept it, implement it, and consequently produce better products. By having incorporated the key components of successful design methodologies, the **Ergodesign Methodology for Product Design** has all the ingredients to turn it into good design practice.

The validity and acceptability of the **Ergodesign Methodology for Product Design** can only be confirmed after its full and effective implementation. This will require that the companies which use it changing their production line to support the design process properly in a mass-production context. Only then can the product and the methodology on which it is based be evaluated against the competition. A preliminary shot at the evaluation problem was to consult some wheelchair designers and to seek their views on the methodology. It was the solution we found since this was originally an academic study and, as such, has its limitations.

The current human-centered methodology was originally created to be used in the design and manufacture of wheelchairs. It is the author's belief that this methodology would be suitable for use for other kinds of products for the disabled or the non-disabled in general. However, due to its level of detail, it may produce better results when used principally for products that have a certain level of complexity. Thus, although the methodology was originally created to be used in a mass-production environment, its use in academia is highly recommended.

So, the **Ergodesign Methodology for Product Design** adapted to each particularity, used with creativity, modern techniques of manufacturing and marketing, and the use of alternative materials (e.g., plastic moldings and thermo-forming components, materials committed to natural and sustainable resources) are essential elements to exploit this potential business opportunity. Marketing research will determine the different market segments for products.

Essentially, the use of this human-centered methodology for product design is an effective option to guarantee that product direct and indirect users have their voice heard and their needs incorporated in a product that will have a good chance of providing full consumer satisfaction.

Appendix 1

Summary of the Ergodesign Methodology for Product Design

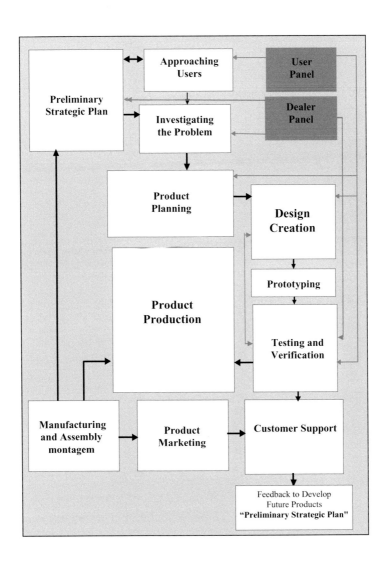

A1.1.   PRELIMINARY STRATEGIC PLAN

Aim:

- Define decision strategies for new product design

[there is little involvement of the designer in this phase].

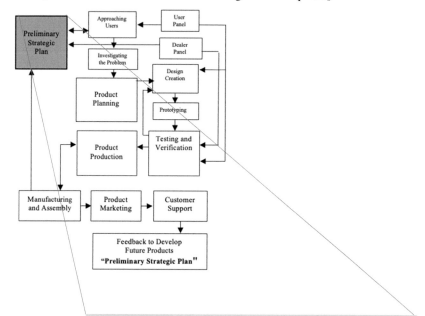

STEPS:
- Define the business planning for the new product, e.g. investigate the business opportunities for the new product.
- Identify the relationship between the new product and the company's other products, e.g. differences between the new product and the competition.
- Define the costs associated with Product Development Program (PDP) expenses.
- Establish a timeline for the PDP.
- Define the preliminary recommendations for innovation.
- Define the applicable technologies.
- Identify target markets.
- Identify competing products and their characteristics.
- Obtain the participation of the dealer Panel.

A1.2. APPROACHING DIRECT AND INDIRECT USERS

Aim:

- Obtain users' opinions about existing products on the market and establish their needs.

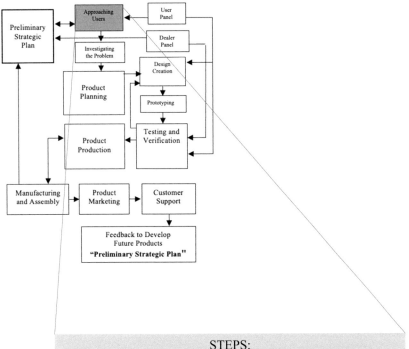

STEPS:
- Investigate existing information about direct and indirect users of the product and its competitors.
- Develop profiles of direct and indirect users.
- Contact direct and indirect users.
- Select direct and indirect users.
- Conduct focus group sessions.
- Select participants for the *User Panel*.

A1.3.   ESTABLISHING THE USER AND DEALER PANEL

Aims:

- Select a group of direct and indirect users [if any] and dealers to participate in the next phases of the design process.
- Involve the team in the design process in order to use their experiences as a source of information for improving the quality and usability of the product.

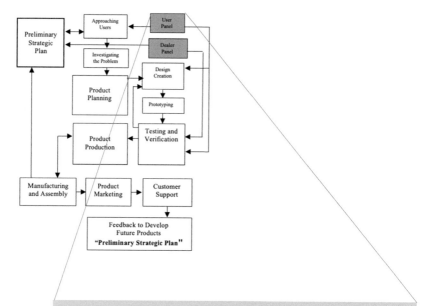

STEPS:

- Provide participants with information about how the product is designed, manufactured and sold including the constraints imposed by the production process.
- Set sessions at significant points in the design process to enable the User Panel and the Dealer Panel, along with the design team, to discuss and make decisions about the next steps in the design process.
- Choose a leader to conduct the sessions.
- Ensure that the User Panel participates in the tasks of analyzing, experimenting with, and evaluating mock-ups, models, prototypes, and instruction manuals.
- Ensure that the Dealer Panel participates in the Preliminary Strategic Planning and Test and Verification stages.

A1.4.   INVESTIGATING THE PROBLEM

Aim:

- Correctly identify the problems to be solved in order to enable the design team to decide what and how to do, considering the competence of the staff, the available knowledge, and what is required by the users.

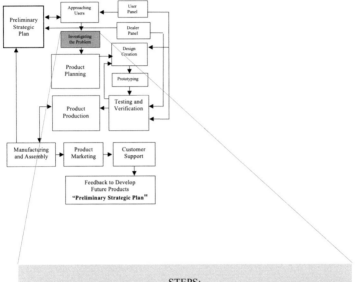

STEPS:

- Describe what is missing in the product or situation being analyzed in terms of meeting the users' needs and/or what exists but does not perform to meet the users' needs.
- Produce a *List of Problems* from the investigation of the most serious problems that arise from the analysis of the situation.
- Select, classify and expand the List of Problems by identifying Ergonomic Dysfunctions [e.g. interfacial and instrumental problems], Human Dysfunctions [e.g. postural and social problems], and Machine Dysfunctions [structural and motion problems].
- Reduce problems to their most significant aspects considering personal skills, available knowledge, and user needs.
- Build the Table Formulation of Problems [Table 1].

A1.4.1 EXAMPLES OF ERGONOMIC PROBLEM DYSFUNCTIONS THAT MAY OCCUR IN THE TABLE FORMULATION OF PROBLEMS

[Product: Wheelchair]

Aim:

- Introduce the most significant solvable problems, design requirements, user limitations, human costs, suggestions for design solutions, and system limitations.

TABLE 1

Problems	Design Requirements	Human Problems	Human costs	Suggestions	Design constraints
Interface					
The backrest does not support the lower back	Backrest profile which considers the buttock protrusion and supports the lumbar region	Dorsal kyphosis and flattening of the lumbar curve	Pain in the back	Provide a new backrest profile	Available technology Lack of interest of buyers and manufacturers
Inappropriate support to accommodate the feet	Considers the length of the feet of the biggest users	Legs do not touch the foot support Pressure in the popliteal region	Discomfort	Provide an adjustable foot support	Lack of interest of buyers and manufacturers
Inappropriate location of push handles	Considers the height of the elbow of the biggest and smaller carers	Flexion of the lumbar spine	Pain in the lower back Pain in the neck	Provide adjustable push handles	Lack of interest of buyers and manufacturers
Control					
Inappropriate shape of the hand controls	Profile that does not cause pressure on the user's hands	Pressure on specific areas of the hands Ulnar/radial deviation	Pain in hands and wrist	Provide new profile for the hand controls	Lack of interest of buyers and manufacturers

A1.5.   PRODUCT PLANNING

Aim:

• Research relevant information for the design team's activities in order to generate and select possible solutions for the creation of new alternatives for the product.

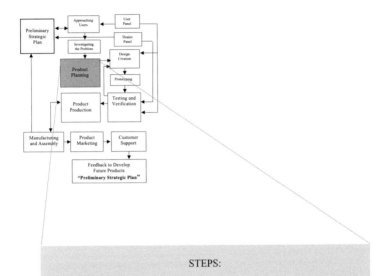

STEPS:

• Perform the Task Analysis to get details about: a) the sequence in which users use the product; b) the hierarchical order of each activity; c) the user-product interface requirements; d) the evaluation and decision of what should be included in the new design; e) the definition of the user's task execution time; and f) the environmental conditions in which the tasks are executed.
• Review the state of the art by: a) reviewing the literature and technical standards;
• Refining the users' needs, translating them into the design process through Product Requirements.
• Build a *List of Refined User Needs and Associated Metrics* [Table 2] based on the Table Formulation of Problems [Table 1].
• Formulate design specifications by analyzing and evaluating competitors' products based on metrics [Table 3] and user satisfaction [Table 4].
• Apply QFD [*Quality Function Deployment*] [Matrix 1].
• Prepare the *Design Specification Document*.

A1.5.1 EXAMPLE OF A PARTIALLY COMPLETE TABLE OF REFINED
USER NEEDS AND THEIR ASSOCIATED METRICS

[Product: Wheelchair]

Aims:

- Specify, in a precise and measurable way, what the product has to do to meet the users' needs.
- Select, categorize and order the importance of each user need that is within the designer's competence to satisfy the user and associate each need with the corresponding measure.

TABLE 2

No.	Subsystem	Needs	Imp.	Metrics	Unity
1	Structure	Reduce the weight of the wheelchair	5	Total mass	kg
2	Structure	Produce foldable wheelchair	4	Fold width	mm
3	Structure	Reduce vibration in the handles	3	Attenuation from push bar to the main structure at 10 Hz	dB
4	Structure	Allow easy traversal of difficult terrain	4	Spring preload	N
5	Structure	Easy to remove wheels	1	Time to disassemble/assemble	min
6	Structure	A wide variety of wheels and tires fit the wheelchair	2	Headset size	mm
				Steer tube diameter	mm
				Wheel sizes	mm
				Castor sizes	mm
				Maximum tire width	mm
7	Structure	Easy access to maintenance of the components	2	Time to disassemble/assemble	min
8	Structure	Easy to maneuver	4	Minimum corridor width of 1000 mm	mm
9	Structure	Is not expensive	5	Unit manufacturing costs	£

A1.5.2 Example of a Partially Complete Table with Competitive Products based on Metrics

[Product: Wheelchair]

Aims:

- Use the data from the List of *Refined User Needs Associated with Metrics* [Table 2] to make a comparison between competitors' products.
- Determine the strengths and weaknesses of competing products relative to the company's products.
- Clarify problems associated with existing products that should be improved in order to increase the chances of success of the product to be designed.

TABLE 3

Metric No.	Need No.	Metrics	Imp.	Unity	Companies				
					A	B	C	D	E
1	1	Total mass	5	kg	15.5	20.0	17.3	16.8	18.0
2	2	Fold width	4	mm	330	580	910	730	815
3	3	Attenuation from push bar to the main structure at 10 Hz	3	dB	12	15	14	12	15
4	4	Preload on the suspension spring	4	N	480	760	500	520	680
5	5	Time to disassemble/ assemble wheels	1	min/ seg	15m 18s	38m 40s	27m 45s	32m 20s	35m 55s
6	6	Headset sizes	2	mm	1.000 1.125	1.000 1.000	1.000 1.250	1.125 1.250	1.125 1.250
7	6	Steer tube diameter	2	mm	254	254	254	254	254
8	6	Wheel size	2	mm	609	558	609	508	628
10	6	Maximum tire width	2	mm	38	44	44	44	44
11	15	Unit manufacturing costs	5	£	1675	1954	1825	2200	2650

A1.5.3 EXAMPLE OF A PARTIALLY COMPLETE TABLE WITH COMPETITIVE PRODUCTS BASED ON USER SATISFACTION

[Product: Wheelchair]

Aims:

- Use the data from the List of the *User Needs Associated with Metrics* [Table 2] to make a comparison between competitors' products based on users' satisfaction and the degree to which different products satisfy their needs.
- Determine the strengths and weaknesses of competing products relative to the company's products in terms of user satisfaction.

TABLE 4

Need No.	Needs	Imp.	Companies				
			A	B	C	D	E
1	Reduce the weight of the wheelchair	5	4	1	3	1	2
2	Produce foldable wheelchair	4	3	2	1	1	1
3	Reduce vibration in the handles	3	2	1	2	2	1
4	Allow easy traversal of difficult terrain	4	1	2	3	1	1
5	Easy to remove wheels	1	3	2	1	1	2
6	A wide variety of wheels and tires fit the wheelchair	2	1	2	3	1	2
7	Easy access to maintenance of the components	2	3	2	1	3	2
8	Sharp edges smoothed off	3	2	1	4	2	1
9	Easy to fit accessories	3	3	2	3	2	2
10	Easy to maneuver	4	4	2	3	2	1
11	Lasts a long time	4	2	3	3	4	4
12	Is safe	5	3	3	3	4	4
13	Is not expensive	5	3	2	1	1	1

A1.5.4   Example of a Partially Complete Matrix with the Application of QFD [Quality Function Deployment]

[Product: Wheelchair]

Aims:

- Translate the users' needs into the product's technical characteristics.
- Identify and quantify the product development steps associated with user needs.
- Translate user demands into design goals and ensure that those demands are used throughout the production stage.

MATRIX 1

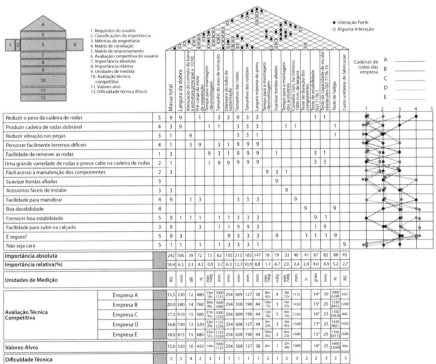

A1.6.   DESIGN CREATION

Aim:

- Generating alternative solutions of a new product in order to satisfy the users' needs, exploiting the potential of sales, marketing and distribution channels, fitting them to the existing manufacturing and supply facilities and finalizing in the generation of profits for the company.

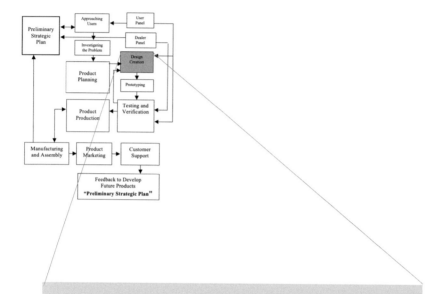

STEPS:

- Generate concepts/alternatives based on the *Product Specification Document*.
- Use special techniques for generating ideas such as brainstroming, brainwriting and synetics.
- Evaluate and select concepts using the *Matrix for Evaluation and Selecting Concepts* [Table 5].
- Refine the concepts using the *Matrix for Refinement of Concepts* [Table 6] to select one or more alternatives capable of being developed.
- Detail the chosen concept in order to build the prototype and proceed to manufacturing.
- Design the User Manual and Promotional Material.

A1.6.1. EXAMPLE OF THE *MATRIX FOR EVALUATING AND SELECTING CONCEPTS*

AIMS:

- Help the design team, with input from the User Panel, to evaluate, compare, select and eliminate different design alternatives.
- Produce new design alternatives from the combination of different features of the various concepts generated.
- Analyze different aspects of the product [such as its subsystems, components, or the combination of these] in terms of their different characteristics, e.g. aesthetics, stability, adjustability, maneuverability, and safety.

TABLE 5

Subsystem: Pushandle					Concepts		
Selection criteria	A	B	C	D	E	F	G
Ease of handling		+	0	0	0	0	0
Ease of use	R	+	0	0	+	0	0
Ease of removal	E	0	0	0	0	+	0
Maneuverability	F	0	-	-	0	+	-
Sharp edges are smoothed off	E	0	0	0	0	0	0
Reduction of vibration in the hands	R E	-	0	0	0	+	-
Good stability	N	0	-	0	+	+	0
Adjustability	C	+	0	+	+	-	0
Help in curb climbing	E	0	0	0	+	0	0
Easy to fit accessories		+	0	0	-	-	0
Wheelchair foldable	C	0	0	+	-	0	0
Safety	O	0	-	0	0	0	0
Low manufacturing costs	N	0	0	0	-	0	+
Sum +'s	C	4	0	2	4	4	1
Sum 0's	E	8	10	10	6	7	10
Sum -'s	P	1	3	1	2	2	2
Net Score	T	3	-3	1	2	2	-1
Rank		1	5	3	2	2	4
Continue?		Yes	No	Yes	Combine	Combine	No

A1.6.2. Example of the *Matrix of Refining Concepts*

Aims:

- Assist the design team, with input from the User Panel, in selecting one or more design alternatives to be produced.
- Analyze different aspects of the product [such as its subsystems, components, or the combination of these] in terms of their different characteristics, e.g. aesthetics, stability, adjustability, maneuverability, and safety.

TABLE 6

Subsystem: Pushandle

Selection criteria	Weight	A R	A Nota	B escore	D Nota	D Esore	EF Nota	EF Escore
Ease of handling	5	E	3	15	2	10	4	20
Ease of use	5	F	2	10	2	10	3	15
Ease of removal	5	E	1	5	1	5	3	15
Maneuverability	10	R	3	30	2	20	4	40
Sharp edges are smoothed off	5	N	3	15	3	15	3	15
		C						
Reduction of vibration in the hands	5	E	3	15	3	15	3	15
Good stability	10		4	40	3	30	4	40
Adjustability	10	C	3	30	2	20	5	50
Help in curb climbing	5	O	4	20	1	5	4	20
Easy to fit accessories	5	N	3	15	3	15	4	20
Wheelchair foldable	5	C	3	15	3	15	4	20
Safety	15	E	3	45	3	45	5	75
Low manufacturing costs	15	P	1	15	3	45	5	75
Total Score		T		270		250		420
Rank				2		3		1
Continue?				No		No		Develop

A1.7. PROTOTYPING

Aim:

- Assist the design team in evaluating product performance so that previously established specifications and user needs are met.

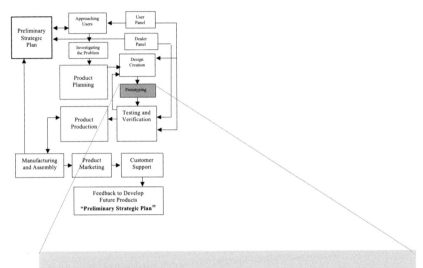

STEPS:

- Build the prototype.
- Analyze whether the concepts represented by the prototype will work and meet both the users' needs and the product's specifications.
- With the help of the User Panel, test the prototype in terms of the user-product interface [including, where appropriate, unpacking and assembly].
- Check that the legal and safety aspects are satisfactory.
- Ensure that components and spare parts are available on the market.
- Check that the costs and production schedule are within the specified limits.

A1.8.   TESTING AND VERIFICATION

Aims:

- Carry out usability tests, with the participation of the User Panel, and compare the result with the product specification data.
- Reduce the likelihood of legal claims against the manufacturer and designer(s).
- Contribute to the success of the product on the market.

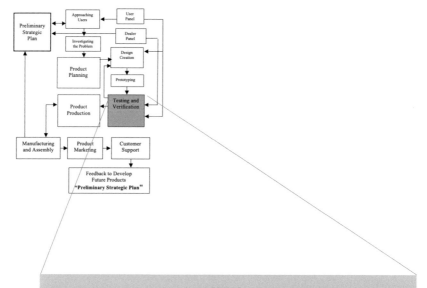

STEPS:

- Provide the conditions for testing.
- Define the resources available in terms of people, equipment, time, money, etc.
- Select the methods and tools used in usability and user experience evaluation
- Establish the test objectives [including what will be measured and why].
- Select the tasks that users will perform.
- Establish the objective and subjective measurements for performance measurement.
- Define the duration of the tests.
- Define the observation and recording techniques that will be used.
- Conduct usability tests, using prototypes, with members of the User Panel and other representative users.
- Analyze and interpret the results making it clear whether or not the prototype is meeting the users' needs.

A1.9. PRODUCT PRODUCTION

Aim:

- Establish the phases involving the product development cycle - assembly methods, manufacturing process, and material selection - based on the details obtained from the previous phases of the design process.

[This step does not directly involve the participation of the designers, so it is not addressed in this methodology.]

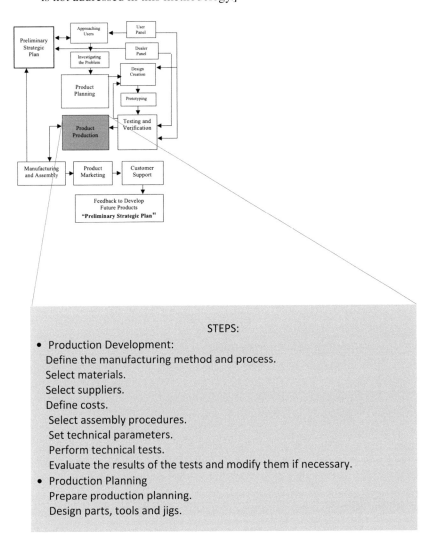

STEPS:
- Production Development:
  Define the manufacturing method and process.
  Select materials.
  Select suppliers.
  Define costs.
  Select assembly procedures.
  Set technical parameters.
  Perform technical tests.
  Evaluate the results of the tests and modify them if necessary.
- Production Planning
  Prepare production planning.
  Design parts, tools and jigs.

A1.10.   MANUFACTURING AND ASSEMBLY

Aim:

• Transform raw materials into manufactured products according to technical specifications.

[This step does not directly involve the participation of the designers, so it is not addressed in this methodology.]

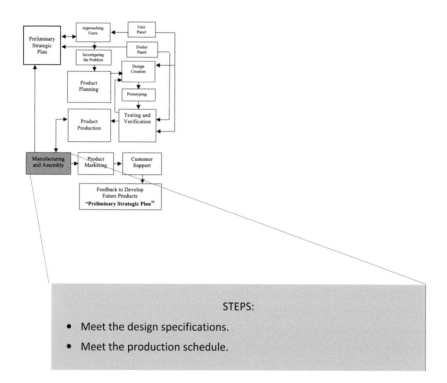

STEPS:

• Meet the design specifications.
• Meet the production schedule.

A1.11. PRODUCT MARKETING

Aim:

- Use marketing and other techniques to advertise and sell the new product.

[This step does not directly involve the participation of the designers, so it is not addressed in this methodology.]

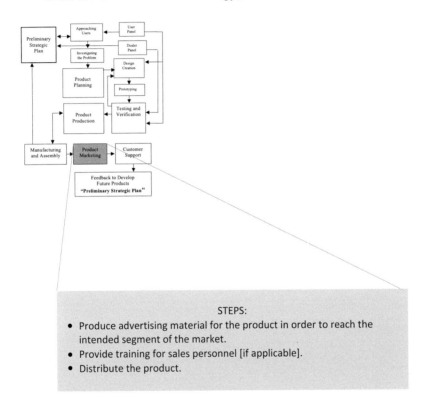

STEPS:
- Produce advertising material for the product in order to reach the intended segment of the market.
- Provide training for sales personnel [if applicable].
- Distribute the product.

A1.12.   CUSTOMER SUPPORT

Steps:

- Provide good technical support in terms of information after purchase, maintenance and repair.
- Monitor the performance of the product while it is being used by the consumer.

[This step does not directly involve the participation of the designers, so it is not addressed in this methodology.]

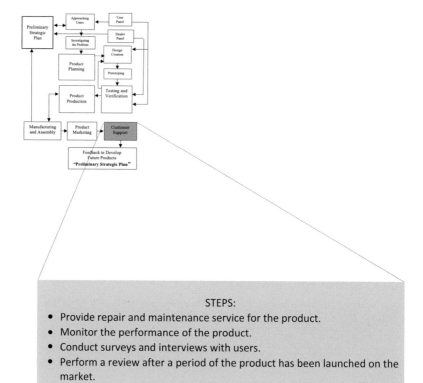

STEPS:
- Provide repair and maintenance service for the product.
- Monitor the performance of the product.
- Conduct surveys and interviews with users.
- Perform a review after a period of the product has been launched on the market.
- Modify the product if necessary.

A1.13. FEEDBACK TO DEVELOP FUTURE PRODUCTS

Aim:

- Evaluate the product's performance in the marketplace and incorporate the results of lessons learned into the planning of future products to be developed.

[This step does not directly involve the participation of the designers, so it is not addressed in this methodology.]

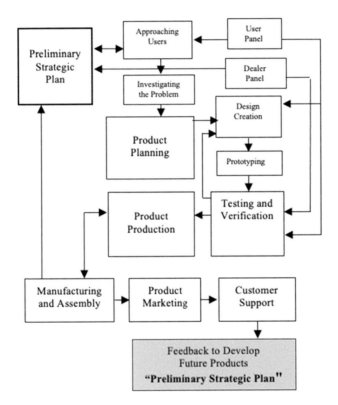

Appendix 2

Major Features of the Results of the Survey of Designers

- Almost all the designers who participated in the survey carried out the design process based on their assumptions about users' expectations. The majority of the wheelchair design processes found in the sample can be considered traditional. Furthermore, by not using systematic methodology, predictions of the product's usage and performance may not match users' expectations.
- The survey has revealed that several phases of the design methodology carried out by the respondents vary significantly from one company to another: some are systematic, some not. The vast majority of design practitioners in the survey do not have an appropriate background involving industrial design qualifications.
- Although respondents regard ergonomics as important in contributing to wheelchair design, its truly effective use in the wheelchair development process has yet to happen.
- Two broad types of error were identified in the analysis of the design process: "errors of omission" and "errors of commission".
- Generally speaking, the main errors of omission were (a) lack of a systematic approach in the design process and (b) not considering the users' requirements in the various design phases. Errors of omission were also identified when small companies that took care of very severely disabled people did not try to overcome communication problems by involving other professionals in the health area.
- Within the topic of design specifications, some companies failed to carry out part or whole phases of the design specification – such as identifying users' needs, evaluating competitive products, establishing user profiles, defining product performance requirements, and determining design constraints.
- It was found that, unfortunately, anthropometric data available in the literature to define the body sizes and shapes of disabled people is almost non-existent. Designers are destined to fail even if they are actively searching for this information. This is one of the reasons given by the respondents for the low use of information from the ergonomics literature.

- The main error of commission identified in the analysis was the fact that managers, technical personnel, and designers made decisions without any involvement of users.
- Consulting only technical personnel on behalf of users with slightly or severely limited communication abilities may be considered an error of commission in the design process of small companies. The ideal process would involve the designer, technical personnel, and health professional to overcome communication problems. Of course, there may be some situations in which this kind of approach would be difficult to adopt.
- Very few manufacturers in the survey were involved in the process of developing and producing entirely new wheelchairs. Broadly speaking, most of the manufacturers preferred to redesign existing wheelchairs rather than to design new products.
- All the respondents stressed the importance of costs in the design process.

Appendix 3

Major Features of the Results of the Survey of Therapists

- Most of the therapists in the sample thought that ergonomics is important in the design of wheelchairs in helping users: (a) to achieve a high level of functional efficiency, conserving energy, and minimizing effort, (b) to ensure that the characteristics of the wheelchair meet individual needs and lifestyle, and (c) to improve posture, movement, and comfort for both users and carers.
- The majority of the respondents identified weaknesses in the process by which clients are assessed, and wheelchair prescribed such as budgetary constraints do not permit clients to be given what they ideally require, or limit the range of wheelchairs available for a prescription; limitation of equipment provided by the statutory service; the user's condition may change during a long waiting time between prescription and delivery; standard wheelchair does not fit client's needs; carers are not considered in the process; the more sophisticated systems are out of the price range of the majority of users, and poor design of wheelchairs causes rejection by the users.
- They stated that the weaknesses in the process by which clients were assessed and wheelchairs prescribed had some implications for the design of wheelchairs, such as the lack of adaptability, interchangeability, and adjustability in a number of wheelchairs available.
- A few more than half of the therapists in the sample said that they did not formally collect the views of the users about the wheelchairs which had been prescribed for them when they have been delivered.
- Amongst those respondents who had collected the views of users about the wheelchairs which had been prescribed, the vast majority of them stated that these views were fed back to designers and manufacturers, for instance, the difficulty the users had in the use of headrest, the need to improve the armrest design, the lack of adjustability of various components.
- The majority of therapists in the sample said that they had, at least once, been in contact with manufacturers about problems connected with wheelchairs. Two-third of them answered that they were either unsure or certain that the manufacturers did not take notice of what they said or that they had, as a consequence, carried out any modification in the wheelchair.

- The majority of them said that although they had never been involved in wheelchair design with a company that mass-produced wheelchairs for a large market, they would like to be involved. They said that they could provide contributions such as reporting experience of clients' needs in their everyday use, home, and workplace; providing feedback from users comments and problems; specifying clinical needs such as activity analysis, functional abilities, posture, seating function; and commenting on technical issues and design features.
- Almost one-third of the respondents thought that, in general, the wheelchairs actually in the marketplace were not designed to take the range of needs of disabled people into account. They argued that the wheelchairs available are: expensive, heavy and bulky, old-fashioned and unattractive, dimensional incompatible, not user-friendly, do not use the technology currently available, and have limited features, variety, and flexibility.

Appendix 4
Major Features of the Results of the Survey of Rehabilitation Engineers

- The respondents pointed out that some weaknesses in the process by which clients were assessed and wheelchairs prescribed had some implications for the design of wheelchairs, such as the lack of adaptability, interchangeability, and adjustability in a number of wheelchairs available.
- Almost half of the sample of rehabilitation engineers answered that they did not formally collect the views of the users about the wheelchairs which had been prescribed for them after delivery.
- Although the communication between rehabilitation engineers and manufacturers has already been established, most of the respondents were not at that stage aware of the need to report users' views to the designers and manufacturers.
- One-third of respondents affirmed that the manufacturers did not take any notice or were unsure if the manufacturers took any notice of what they said and consequently carried out any modification to the wheelchairs.
- The vast majority of respondents had never been involved in wheelchair design with a company that produces wheelchairs for a large market but would like to be involved. They thought that their main contributions might include: providing information gained from practical experience with users and their requirements; providing technical contribution, including seating, posture management, and ergonomics; providing feedback on the problems users have, including design solutions; advising on design suitability; and taking part in product evaluation and trials of prototypes.
- Almost half of the respondents thought that, in general, the wheelchairs actually in the marketplace were not designed to take the range of needs of disabled people into account. They argued that there is a lack of knowledge about users' needs or the disabled are not asked for their opinions; the wheelchairs are generally designed for those who are least demanding or to suit an average person; the price of the chairs is high; there is a lack of adjustability in standard models; there are not enough field trials to iron out the design faults; they are only aimed at the young active user; they are very heavy; and few of the wheelchairs available are crash-tested.

Appendix 5

Major Features of the Results of the Survey of Users

The majority of wheelchair users in the survey:

- were over 45 years old, and more than one-third of them were over 55 years old;
- suffered from neurological conditions;
- lived in an urban area: town or city;
- took their wheelchair with them when they went out in a vehicle;
- had used some form of public transport in the last 12 months, such as airplanes, intercity, and local train, taxi, and low floor bus;
- had two or more wheelchairs;
- had been using a wheelchair for more than ten years;
- owned a manual self-propelled wheelchair as the most-used and the next most-used wheelchair;
- obtained their wheelchairs through the U.K. Governmental Agency, for both the most-used and the next most-used wheelchair;
- had owned their current wheelchair(s) for less than five years;
- had a seat cushion for both wheelchairs (the most-used and the next most-used wheelchair);
- used their main wheelchair every day;
- used their main wheelchair indoors for more than five hours a day;
- had had problems with their main wheelchair in the last 12 months such as punctures, electrical failures, brake failure, or broken armrest;
- considered safety, comfort, reliability, suitability, and portability due to weight as the five most important design characteristics of wheelchairs;
- judged the design of their own wheelchairs as being "very good" or "good" in terms of safety, ease of use, stability, maneuverability, suitability, and reliability, although this level of satisfaction was not achieved as consistently as might be hoped for;
- judged the design of their own wheelchairs as being "average", "poor", or "very poor" in terms of cost to buy, cost to repair, provision of accessories, cost to maintain, adjustability, ease of repair, aesthetic appearance, and portability due to weight;

- had the view that privately acquired wheelchairs were designed taking into consideration the needs of disabled people and those issued by the U.K. Governmental Agency were not; and
- had never been involved in wheelchair design with a company that mass-produced wheelchairs for a large market.

Appendix 6

Major Features of the Results of the Survey of Carers

The majority of carers in the survey:

- were over 35 years old, and almost one-quarter of them were over 65 years old;
- assisted users in using the wheelchair every day;
- rated their own health at the time they answered the questionnaire as "average", "poor", or "very poor";
- suffered from pain in the regions of lower back, buttocks, mid-back, and right shoulder as a consequence of assisting the user with the wheelchair;
- considered safety and portability due to weight as the most important design characteristics of wheelchairs;
- judged the design of the wheelchairs belonging to the wheelchair user who they assisted as being "very good" or "good" in terms of safety, ease of use, stability, reliability, robustness, and suitability, although this level of satisfaction was not achieved as consistently as might be hoped for;
- judged the design of the wheelchairs belonging to the wheelchair user whom they assisted as being "average", "poor", or "very poor" in terms of cost to buy, cost to repair, provision of accessories, cost to maintain, and aesthetic appearance;
- had the view that the wheelchairs issued by the U.K. Governmental Agency were not designed to take into consideration the needs of disabled people and their carers; and
- had never been involved in wheelchair design with a company that mass-produced wheelchairs for a large market.

Appendix 7

Consulting Designers to Confirm the Suitability of the Proposed Methodology

A sample of four designers who had previously participated in the field study at the start of the work was approached. This relatively small number was due to the time available, which made it impossible to have more respondents involved. The aim was to collect their views on the extent to which the proposed methodology was acceptable to them. The companies the designers worked for were all located in the UK.

The main criterion for choosing the selected designers was to represent those wheelchair companies that provided the best practice in terms of a questionnaire on Design Methods applied by the author in his Ph.D. thesis. One of these companies was considered at the time as one of the biggest wheelchair manufacturers in the United Kingdom. Another company was a manufacturer of scooters, whose design and production process has a number of similarities with that of electrical wheelchairs. The choice of wheelchair designer was because the methodology was originally developed for the design of wheelchairs. Since the methodology retains its original essence, the validation of the designers remains appropriate. Below, we will present a summary of this consultation.

1. PROCEDURES

Each designer was willing to be interviewed in their workplace for about one hour. The designers were shown a summary of the methodology in a form to be read and commented on during the time available for the interview. It was decided not to structure the interview with specific questions to avoid guiding the interviewee and drawing attention to certain points. Instead, they were asked to read the summary and to produce comments while they were reading the text. In this way the researcher tried to explore their comments, questioning respondents when appropriate. The interviews were tape-recorded and transcribed. The more relevant comments provided by the designers are described below.

Generally speaking, the designers who investigated the suitability of the proposed methodology for wheelchair design provided positive comments. There was not a pre-presentation of the methodology, and it was found to be difficult to introduce a methodology with such a level of complexity in only one hour of interview. Although a file with a graphical summary of the methodology was used, the respondents sometimes made comments on aspects of design that were already taken into consideration, but which were not described in detail in the summary form presented to the designers. For example, the designers said that more people should be used than the number prescribed for the **User Panel**. However, this would occur in the focus group sessions in the phase concerned with **Approaching the Users**.

2. MOST RELEVANT COMMENTS

- The involvement of other stakeholders (e.g., therapists and rehabilitation engineers) in the design process, suggested by the designer of Company A, and the needs for the **User Panel** to be representative of the different ranges of disabilities had already been taken into consideration. The suggestion to involve dealers as part of the design process was an important point and was incorporated into the new version of the methodology. The involvement of the largest number of people (actors) interested in the product was also suggested as an important factor when producing products to serve the consumer market.
- The importance of considering standards in the design methodology, mentioned by two designers. However, this consideration is part of the "Reviewing the state of the art", in the phase of **Product Planning** but was apparently overlooked.
- Correctly interpreting the needs of the users, as stated by one of the designers, is without doubt one of the points responsible for the success of a product. It will basically depend on the skill of the design team. The help of other stakeholders, such as carers and therapists, may be useful to help to clarify points which the users have difficulty in articulating themselves.
- A loop in the phases of **Design Creation**, **Prototype**, and **Testing and Verification** should be considered in such a way that the concept that is not successful in the phase of **Testing and Verification** may be returned to be modified in the **Design Creation** and **Prototyping** phases again.
- The need to assess the phase of **Manufacturing and Assembly** as a function of what was defined in the **Preliminary Strategic Planning** should be taken into consideration.
- The possibility of using the **Customer Support** phase to monitor the product performance in the marketplace and obtain feedback to be used for other products to be developed by the company in the future should also be considered.
- The designers were unanimous about the quality of the methodology. They drew attention (with the exception of the designer of Company C) to the fact that, in their opinion, this methodology is only justified if it is applied to manufacturing on a large scale.

References

Abbott, H. (1980). *Safe Enough to Sell? Design and Product Liability*. London, The Design Council. ISBN-13: 978-0850721133.

Abbott, H.; Tyler, M. (2017). *Safer by Design: A Guide to the Management and Law of Designing for Product Safety*, 2nd ed. London, The Design Council.

Abeni, K. (1988). An assessment of industrial designers use of human factors criteria in product design evaluation. In: Proceedings of the Human Factors Society–32nd Annual Meeting. Santa Monica, CA, Human Factors Society, 420–424.

Acioly, A. (2016). *Augmented Reality as a Tool to Guide Use and Safety in Packaging (A realidade aumentada como ferramenta para orientação de uso e de segurança em embalagens)*. In Portuguese. Ph.D. Thesis. Post Graduate Program in Design. Federal University of Pernambuco, Brazil.

Ahram, T.; Barros, R.Q.; Falcão, C.; Soares, M.M.; Karwowski, W. (2016). Neurodesign: Applications of neuroscience to design and human-system interactions. In: Marcelo M. Soares; Francisco Rebelo. (Org.). *Ergonomics in Design: Methods and Techniques*, 1st ed. Boca Raton, Estados Unidos: CRC Press, 481–496. ISBN-13: 978-1498760706.

Ahram, T.; Karwowski, W.; Soares, M.M. (2011). Smarter products user-centered systems engineering. In: Karwowski, W.; Soares, M.M.; Stanton, N.A. (Org.). *Human Factors and Ergonomics in Consumer Product Design: Methods and Techniques*, vol. 1. Boca Raton, FL: CRC Press, 83–94. ISBN-13: 978-1420046281.

Akao, Y. (2004). *QFD: Quality Function Deployment: Integrating Customer Requirements into Product Design*. Productivity Press. ISBN-13: 978-1563273131. New York

Albert, W.; Tullis, T. (2013). *Measuring the User Experience: Collecting, Analyzing, and Presenting Usability Metrics*, 2nd ed. Burlington, MA: Morgan Kaufmann. ISBN-13: 978-0124157811.

American Society for Quality-ASQ (2019). *What is Quality Function Deployment?* Retrieved August 25th 2020 from https://asq.org/quality-resources/qfd-quality-function-deployment.

Anderson, D. (2014). *Design for Manufacturability: How to Use Concurrent Engineering to Rapidly Develop Low-Cost, High-Quality Products for Lean Production*. Boca Raton, FL: CRC Press. ISBN-13: 978-1482204926-.

Andre, A.D.; Segal, L. (1994). Design function. *Ergonomics in Design*, October, 6–8, 1, 4–5. DOI: 10.1177/106480469300100102.

Annet, J.; Stanton, N.A. (2000). *Task Analysis*. Boca Raton, FL: CRC Press. ISBN-13: 978-0748409068.

Arora, A. et al. Problematic use of digital technologies and its impact on mental health during COVID-19 pandemic: Assessment using machine learning. *Emerging Technologies during the Era of COVID-19 Pandemic*, 348, 197–221. 21 March 2021, Available at: https://www.ncbi.nlm.nih.gov/pmc/articles/PMC7980158/

ASQ (2020). *What is Quality Function Deployment (QFD)*. Retrieved September 16th 2020 from https://asq.org/quality-resources/qfd-quality-function-deployment.

Baber, C.; Neville A. Stanton (2002) Task analysis for error identification: Theory, method and validation. *Theoretical Issues in Ergonomics Science*, 3(2), 212–227. DOI: 10.1080/14639220210124094.

Barber, J. (1996). The design of disability products: A psychological perspective. *British Journal of Occupational Therapy*, 59(12), 561–564. DOI: 10.1177/030802269605901205.

Barros, R.Q. (2016). *Application of Neuroergonomics, Eye Tracking and Infrared Thermography in the Evaluation of Consumer Products: A Usability Study (Aplicação da Neuroergonomia, Rastreamento Ocular e Termografia por Infravermelho na avaliação de produto de consumo: um estudo de usabilidade).* In Portuguese. M.Sc. thesis. Post Graduate Program in Design, Federal University of Pernambuco, Brazil.

Barros R.Q. et al. (2016) Analysis of product use by means of eye tracking and EEG: A study of neuroergonomics. In: Marcus A. (ed.). *Design, User Experience, and Usability: Novel User Experiences. DUXU 2016. Lecture Notes in Computer Science*, vol. 9747. Cham: Springer. https://doi.org/10.1007/978-3-319-40355-7_51

Barros R.Q.; Santos G.; Ribeiro C.; Torres R.; Barros M.Q.; Soares M.M. (2015). A usability study of a brain-computer interface apparatus: An ergonomic approach. In: Marcus A. (ed.). *Design, User Experience, and Usability: Design Discourse. Lecture Notes in Computer Science*, vol 9186. Cham: Springer. https://doi.org/10.1007/978-3-319-20886-2_22

Baxter, M. (1995). *Product Design: A Practical Guide to Systematic Methods of New Product Development.* Boca Raton, FL: CRC Press. ISBN-13: 978-1138442863.

Betsky, A. (2021). Design in a Post-COVID-19 World. *The Journal of the American Institute of Architects.* Available at: https://www.architectmagazine.com/design/design-in-a-post-covid-19-world_o Accessed on: May15th 2021.

Blackwell, R.D.; Miniard, P.W.; Engel, J.F. (2005). *Consumer Behavior.* 10th ed. Mason, OH: South-Western College Pub. ISBN-10: 9780324271973.

Brand, et al. (2014). Prefrontal control and internet addiction: A theoretical model and review of neuropsychological and neuroimaging findings. *Frontiers in Human Neuroscience*, 8, 375.

Brangier, E.; Bornet, C. (2011). Persona: A method to produce representations focused on consumers' needs. In: Karwowski, W.; Soares, M.M. and Stanton, N. (eds.). *Human Factors and Ergonomics in Consumer Product Design: Methods and Techniques.* Boca Raton, FL: CRC Press. ISBN-13: 978-1420046281.

Bridger, R. (2017). *Introduction to Human Factors and Ergonomics*, 4th ed. Boca Raton, FL: CRC Press. ISBN-13: 978-1498795944.

Brown, G.N.; Wier, A.P. (1982). Human factors and industrial design (are we really working together?). In: Proceedings of the Third National Symposium on Human Factors and Industrial Design in Consumer Products. Santa Monica, CA, Human Factors Society, 3–10.

Bruseberg, A.; McDonagh-Philp, Deana. (2002). Focus groups to support the industrial/product designer: a review based on current literature and designers' feedback. *Applied Ergonomics*, 3(1), 27–38. ISSN 0003-6870.

BS 7000–6 (2005). *Design Management Systems: Managing Inclusive Design.* London: British Standards Institute.

Canadian Centre for Occupational Health and Safety (2019). *Hazard and Risk.* Retrieved July 3rd 2019 from https://www.ccohs.ca/oshanswers/hsprograms/hazard_risk.html.

Caplan, S. (1990). Using focus group methodology for ergonomic design. *Ergonomics*, 33(5), 527–533.

Casali, E.F. (2021). Co-design and participatory design: A solid process primer. *Intense Minimalist.* Available at: https://intenseminimalism.com/2013/co-design-and-participatory-design-a-solid-process-primer/. Accessed on: September 10th 2021.

Casey, S.M. (1998). *Set Phasers on Stun: And Other True Tales of Design, Technology, and Human Error.* Santa Barbara, CA: Aegean Pub Co. ISBN-10: 9780963617880.

Cavignau-Bros, D.; Cristal, D. (2020). Participatory design and co-design: The case of MOOC on public innovation. In: Schmidt, M.; Tawfik, A. A.; Jahnke, I.; Earnshaw, Y. (eds.). *Learner and User Experience Research: An Introduction for the Field of*

Learning Design & Technology. EdTech Books. Available at: https://edtechbooks.org/ux/participatory_and_co_design. Accessed on: September 10th 2021.

Chaffin, D.B.; Andersson, G.B.J.; Martin, B.J. (2006). *Occupational Biomechanics*, 4th ed. New York: Wiley-Interscience. ISBN-13: 978-0471723431.

Chang, K. (2016). *Computer-Aided Engineering Design*. Cambridge, MA: Academic Press. ISBN-13: 978-0128095690.

Chapanis, A. (1996). *Human Factors in Systems Engineering*. New York: Wiley. ISBN-13: 978-0471137825.

Chapman, J. (2015). *Emotionally Durable Design: Objects, Experiences and Empathy*, 2nd ed. London: Routledge. ISBN-13: 978-0415732154.

Chartered Institute of Ergonomics and Human Factors (2019). User Centred Design: A Practical Guide for Teachers. Retrieved September 27th 2019 from https://www.ergonomics.org.uk/Public/Resources/Publications/User_Centred_Design/Public/Resources/Publications/UCD.aspx?hkey=2138360f-c44a-4c20-b500-1efb4f04e43a.

Chartered Institute of Ergonomics and Human Factors (2021). User Centred Design: a Practical Guide for Teachers. Available at: https://www.ergonomics.org.uk/Public/Resources/Publications/User_Centred_Design/Public/Resources/Publications/UCD.aspx?hkey=2138360f-c44a-4c20-b500-1efb4f04e43a. Accessed on: September 27th 2021.

Children's Products (2020). *What Requirements Apply to My Product*. Retrieved November 10th 2020 from https://www.cpsc.gov/Business--Manufacturing/Business-Education/childrens-products.

Christensen, J.M. 1987, Comments on products safety, rising to the new heights with technology. In: Proceedings of the Human Factors Society: 31st Annual Meeting. Human Factors Society, Santa Monica, CA, 1–14.

Christiaans, H.H.C.M. (1989). The use of consumer products: A cognitive view. In: Proctor, G.E.; Stadelmeier, S.; Stubler, W.; Opperman, L. and Kusuma, D. (eds.). Product Design: Facts vs. Feelings, Proceedings of Interface '89. Human Factors Society, Santa Monica, CA, Consumer Products Technical Group of the Human Factors Society, 189.

Churchill, G.A.; Surprenant, C. (1982). An investigation into the determinants of customer satisfaction. *Journal of Marketing Research*, XIX, 491–504.

Cifter, D.H. (2010). Instruction manual usage: A comparison of younger people, older people and people with cognitive disabilities. In: Winter R., Zhao J.L., Aier S. (eds.). *Global Perspectives on Design Science Research. DESRIST 2010. Lecture Notes in Computer Science*, vol 6105. Berlin, Heidelberg: Springer. DOI: 10.1007/978-3-642-13335-0_28.

Clarkson, P.J.; Coleman, R.; Keates, S. (2003). *Inclusive Design: Design for the Whole Population*. Cham: Springer. ISBN-13: 978-1852337001.

Cleverism (2019). *Why Most Product Launches Fail (And What To Do About It)*. Retrieved July 3rd 2019 from https://www.cleverism.com/why-most-product-launches-fail/

Collaborating With Customers in Product Development: CBS News (2007). Retrieved July 10th 2019 from https://www.cbsnews.com/news/collaborating-with-customers-in-product-development/

Consumer Reports (2019). Unsafe by definition: Substantial product hazard. *Consumer Reports News*, September 07, 2010. Retrieved July 3rd 2019 from https://www.consumerreports.org/cro/news/2010/09/unsafe-by-definition-substantial-product-hazard/index.htm

Cooper, R.A.; Robertson, R.N.; Boninger, M.L.; Shimada, S.D.; Van Sickle, D.P.; Lawrence, B.; Singleton, T. (1997). Wheelchair ergonomics. In: Kumar, S. (ed.). *Perspectives in Rehabilitation Ergonomics*. London: Taylor and Francis. ISBN-13: 978-0748406449.

Cross, N. (2008). *Engineering Design Methods: Strategy for Product Design*, 4th ed. New York: Wiley. ISBN-13: 978-0470519264.

Csikszentmihalyi, M.; Larson, R. (2014). Validity and reliability of the experience-sampling method. *The Journal of Nervous and Mental Disease*, 175(9), 35–54. DOI: 10.1097/00005053-198709000-00004.

Cuffaro, D. et al. (2013). *Industrial Design. Reference + Specification Book*. Rockport. ISBN-13: 978–1592538478.

Curedale, R. (2019). *Design Thinking: Process & Methods*, 5th ed. Los Angeles: Design Cummunity College. ISBN-13: 978-1940805450.

Cushman, W.H.; Rosenberg, D.J. (1991). *Human Factors in Product Design*. Amsterdam: Elsevier. ISBN-13: 978-0444874344.

Cybis, W.; Betiol, A.; Faust, R. (2017). *Ergonomia e Usabilidade: Conhecimentos, Métodos e Aplicações*. Sao Paulo: Novatec Editora.

Cyr, J. (2019). *Focus Groups for the Social Science Researcher*. Cambridge: Cambridge University Press. ISBN-13: 978-1316638798.

Dahlin, T.; Mascanzoni, D.; Rosell, G.; Svengren, L. (eds.) (1994). *The Human Dimension: Swedish Industrial Design*. Bergamo, Italy: Edizioni Bolis. ISBN-13: 978-9163028847.

Dam, R.; Siang, T. (2019). *What is Design Thinking and Why Is It So Popular?* Interaction Design Foundation. Retrieved October 17th 2019 from https://www.interaction-design .org/literature/article/what-is-design-thinking-and-why-is-it-so-popular

Damasio, A. (2015). *The Mystery of Conscience (O mistério da consciência)*. In Portuguese. Sào Paulo: Companhia das Letras. ISBN-13: 978-8535925906.

De Feo, J. (ed.) (2017). *Juran's Quality Control Handbook: The Complete Guide to Performance Excellence*, 7th ed. New York: McGraw-Hill. ISBN: 978-1-25-964361-1.

Dekker, S. (2014). *The Field Guide to Understanding Human Error*. 3rd ed. Boca Raton, FL: CRC Press. ISBN-13: 978–0754648260.

Design bei Rollstühlen (1993). Form, Journal for design 93. Special edition for the Group of Product Design, Kastanienallee 20, D-6450 Hanau/M.1 (in German).

Desmet, P. (2002). *Designing Emotions*. Ph.D. Thesis. Delft University of Technology, Delft.

Dewis, M.; Hutchins, D.C.; Madge, P. (1980). *Product Liability*. London: Heinemann. ISBN-13: 978-0434903115.

Dirken, J.M. (2007). Approved by ergonomists? *Ergonomics*, 33, 269–273. 10.1080/00140139008927122.

Diverstiy & Inclusion in Tech: A Practical Guide for Entrepreneurs (2018). Retrieved July 10th 2019 from https://www.inclusionintech.com/wp-content/uploads/2018/12/Divers ity_Inclusion_in_Tech_Guide_2018.pdf

Dumas, J.; Loring, B. (2008). *Moderating Usability Testing: Principles and Practices for Interacting*. Burlington, MA: Morgan Kauffman Publishers. ISBN-13: 978-0123739339.

Dumas, J.S.; Redish, J.C. (1999). *A practical Guide to Usability Testing*. Norwood, NJ: Intellect Ltd. ISBN-13: 978-1841500201.

Eason, K.D. (2007). Towards the experimental study of usability. Behaviour and Information Technology, 3(2), 133–143. DOI: 10.1080/01449298408901744.

Ekman, P.; Rosenberg, E.L. (2020). *What the Face Reveals: Basic and Applied Studies of Spontaneous Expression Using the Facial Action Coding System (FACS)*. 3rd ed. Oxford: Oxford University Press. ISBN-13: 978-0190202941.

Elkind, J. (1990). The incidence of disabilities in the United States. *Human Factors*, 32(4), 397–405. DOI: 10.1080/01449298408901744.

Endsley, M. (2017). *Designing for Situation Awareness: An Approach to User-Centered Design*. 2nd ed. Boca Raton, FL: CRC Press. ISBN-13: 978-1138460416.

Erhorn, C.; Stark, J. (1994). *Competing by Design: Creating Value and Market Advantage in New Product Development*. Essex Junction: Oliver Wight Publications. ISBN-13: 978-0471132165.

Esperidião-Antonio, V. et al. (2008). Neurobiology of emotions (Neurobiologia das emoções). In Portuguese. *Revista de Psiquiatria Clínica*, 35(2), 55–65. DOI: 10.1590/S0101-60832008000200003.

Eye Tracking (2020). Usability.gov. Disponível em https://www.usability.gov/how-to-and-tools/methods/eye-tracking.html. Accessed on: August 14th 2020.

Falcão, C.S.; Soares, M.M. (2013). Usability of consumer products: An analysis of concepts, methods and applications (Usabilidade de produtos de consumo: uma análise dos conceitos, métodos e aplicações). In Portuguese. *Estudos em Design. Rio de Janeiro*, 21(2), 1–26.

FastCompany (2021). The hot fashion accessory of 2020? Masks, masks, and more masks. Available at: https://www.fastcompany.com/90494852/the-hot-fashion-accessories-of-2020-masks-masks-and-more-masks. Accessed on: May 14th 2021.

Feeney, R.J. (1996). Participatory design: Involving users in the design process. In: Özok, A.F. and Salvendy, G. (eds.). Advances in applied ergonomics. Proceedings of the 1st. International Conference on Applied Ergonomics, Istanbul, Turkey, May 21–24. West Lafayette, USA Publishing, 199–203.

Feeney, R.J.; Galer, M.D. (1981). Ergonomics research and the disabled. *Ergonomics*, 24(11), 821–830. DOI: 10.1080/00140138108924903.

Forcelini, F. et al. (2018). Creativity techniques in the design process (As técnicas de criatividade no processo de design). *Temática*, NAMID/UFPB, 14(1), 31–46. Retrieved July 29th 2020 from https://pdfs.semanticscholar.org/86ab/e1c7d16b40b79e2471de77e77d033fc0f7ef.pdf

Fox, J. (1993). *Quality through Design: The Key to Successful Product Delivery*. London: McGraw-Hill.

Gao, J. et al. (2020). Mental health problems and social media exposure during COVID-19 outbreak. *PLoS ONE*, 15(4). doi: 10.1371/journal.pone.0231924

Gardiner, P.; Rothwell, R. (1985). Tough customers: Good design. *Design Studies*, 6, 7–17.

Garvin, D.A. (1988). *Managing Quality: The Strategic and Competitive Edge*. New York and London: The Free Press and Collier Macmillan Publishers. ISBN-13: 978-0029113806.

Gatti, B.A. (2012). Focus Groups in Social Science and Humanities Research *(Grupo Focal na Pesquisa em Ciências Sociais e Humanas)*. In Portuguese. Brasilia: Liber Livro. ISBN-13: 978–8598843117.

Gensler (2021). Design responds to a changing world. Available at: https://www.gensler.com/design-responds-to-a-changing-world. Accessed on: May 15th 2021.

Godman, E.; Kuniavsky, M.; Moed, A. (2012). *Observing the User Experience: A Practitioner's Guide to User Research*. Widham, MA: Morgan Kaufmann. ISBN-13: 978-0123848697.

Goetsch, D.; Davis, S. (2015). *Quality Management for Organizational Excellence: Introduction to Total Quality*, 8th ed., London: Pearson. ISBN-13: 978-0133791853.

Goldberg, J.H.; Wichansky, A.M. (2003). Eye tracking in usability evaluation: A practitioner's guide. In: Radach, R.; Hyona, J.; Deubel, H. (eds.). *The Mind's Eye*. Amsterdam: Elsevier Science, 493–516. DOI:10.1016/B978-044451020-4/50027-X

Gomes Filho, J. (2020). *Object ergonomics: Technical reading systems (Ergonomia do objeto: sistemas técnicos de leitura)*, 2nd ed. In Portuguese. São Paulo: Escrituras. ISBN-13: 978-8575313602.

Goodwin, K. (2009). *Designing for the Digital Age: How to Create Human-Centered Products and Services*. New York: Wiley. ISBN-13: 978-0470229101.

Govella, A. (2019). *Collaborative Product Design*. Sebastopol, Canada: O'Reilly Media. ISBN-13: 978-1491975039.

Grandjean, E. (1984). Ergodesign 84: Ergonomics and design in the electronic office. *Behaviour and Information Technology*, 4(1), 75–76, DOI: 10.1080/01449298508901789

Green, W.; Jordan, P.W. (1999). *Human Factors in Product Design: Current Practice and Future Trends*. Boca Raton, FL: CRC Press. ISBN-13: 978-0748408290.

Green, W.S.; Jordan, P.W. (eds.) (2002). *Pleasure with Products: Beyond Usability*. London: Taylor and Francis. ISBN-13: 978-0415237048.

Gregory, K. (1982). Determining the 'consumer object'. *Applied Ergonomics*, 13, 11–13.

Griffin, A.; Hauser, J.R. (1993). The voice of customer. *Marketing Science*, 12(1) winter, 1–27.

Gruman, G. (2021). How to set up a WFH 'office' for the long term. ComputerWorld. Available at: https://www.computerworld.com/article/3545478/how-to-set-up-a-work-from-home-office-for-the-long-term.html. Accessed on: May 21st 2021.

Gryna, F.M. (2016). Product development. In: Juran, J.M. and Gryna, F.M. (eds.). *Juran's Quality Control Handbook*, 7th ed. New York: McGraw-Hill. ISBN-10: 9781259643613.

Gullo, L.J.; Dixon, J. (2018). *Design for Safety*. New York: Wiley. ISBN-13: 978-1118974292.

Hale, G. (1979). *The Source Book for the Disabled: An Illustrated Guide to Easier and More Independent Living for Physically Disabled People, Their Families, and Friends*. New York: Paddington Press. ISBN-13: 978-044822426.

Hamraie, A. (2017). *Building Access: Universal Design and the Politics of Disability*, 3rd ed. Minneapolis, MI: University of Minnesota Press. ISBN-13: 978-1517901639.

Hanington, B.; Martin, B. (2012). *Universal Methods of Design: 100 Ways to Research Complex Problems, Develop Innovative Ideas, and Design Effective Solutions*. Beverly, MA: Rockport Publishers. ISBN-13: 978-1592537563.

Harker, S.D.P.; Eason, K.D. (1984). Representing the user in the design process. *Design Studies*, 5(2), 79–85. DOI: 10.1016/0142-694X(84)90040-1

Harris, C.M.-T. (1990). A study in the marketing of ergonomic expertise in the industrial setting. *Ergonomics*, 33, 547–552.

Hassenzahl, M.; Tractinsky, N. (2006). User experience: A research agenda. *Behaviour and Information Technology*, 25(2), 91–97. DOI: 10.1080/01449290500330331.

Hassenzahl, M.; Burmester, M.; Koller, F. (2003). AttrakDiff: Ein Fragebogen zur Messung wahrgenommener hedonischer und pragmatischer Qualitat. In J. Ziegler and G. Szwillus (eds.). *Mensch&Computer 2003. Interaktion in Bewegung*. Stuttgart, Leipzig: B. G. Teubner, 187–196.

Hauser, J.R.; Clausing, D. (1988). The House of Quality. Havard Business Review. Available at: https://hbr.org/1988/05/the-house-of-quality. Accessed: April 21st 2021.

Healthline (2021). 26 WFH tips while self-isolating during the COVID-19 outbreak. Available at: https://www.healthline.com/health/working-from-home-tips. Accessed on: May 21st 2021.

Hekstra, A.C. (1993). Human factors in wheelchair testing. In: Woude, C.H.V. van der; Meijs, P.J.M.; Griten, B.A.; van der and Boer, Y.A. de (eds.). *Ergonomics of Manual Wheelchair Propulsion*. Oxford, UK: I.O.S. Press.

Helander, M.; Khalid, H. (2012). Affective engineering and design. Capítulo 20, In: Salvendy, G. (ed.). *Handbook of Human Factors and Ergonomics*, 3rd ed. New York: Wiley. ISBN-13: 978-0470528389.

Hodgson, P. (2019). Tips for writing user manuals. Userfocus, June, 4th 2007. Retrieved September 23rd 2019 from https://www.userfocus.co.uk/articles/usermanuals.html

Holt, K. (1989). Does the engineer forget the user? *Design Studies*, 10, 163–160. DOI: 10.1016/0142-694X(89)90034-3.

Hom, J. (2020). *The Usability Methods Toolbox Handbook*. Retrieved August 13th 2020 from http://usability.jameshom.com.

HSSE World (2021). *What is the difference between a hazard and a risk?* HSSE World – Health, Safety, Security and Environment. Available at: https://hsseworld.com/what-is-the-difference-between-a-hazard-and-a-risk/. Accessed on: May 28th 2021.

Hunter, T.A. (1992). Engineering Design for Safety. New York: McGraw-Hill. ISBN-13: 978-0070313378.

Hunter Jr., R.; Shannon, J.H.; Amoroso, H.J. (2018). *Products Liability: A Managerial Perspective.* Independently published. ISBN-13: 978-1731150684.

ICANotes (2018). How digital addiction affect us. ICANotes Behavioral Health EHR. Available at: https://www.icanotes.com/2018/12/29/how-digital-addiction-affects-us/. Accessed on: May 18th 2021.

IDEO (2015). *The Field Guide to Human-Centered Design.* IDEO.org/Design Kit. ISBN-13: 978-0991406319.

Iida, I.; Guimarães, L. (2016). *Ergonomics: Design and Production (Ergonomia: projeto e produção),* 3rd ed. In Portuguese. Sao Paulo: Edgard Blücher. ISBN-13: 978-8521209331.

ILO (2020). Teleworking during the COVID-19 pandemic and beyond: A practical guide. Available at: https://www.ilo.org/wcmsp5/groups/public/---europe/---ro-geneva/---ilo-lisbon/documents/publication/wcms_771262.pdf. Accessed on: May 17th 2021.

Injury Facts (2019), *Home and Community Overview.* Retrieved June 30th 2019 from https://injuryfacts.nsc.org/home-and-community/home-and-community-overview/introduction/

Interaction Design Foundation (2020). Conducting a Focus Group. Available at: https://www.interaction-design.org/literature/article/how-to-conduct-focus-groups. Accessed on June 1st. 2021.

International Ergonomics Association (IEA) (2020). Retrieved August 4th 2020 from https://iea.cc.

Ioannou, S.; Gallese, V.; Merla, A. (2014). Thermal infrared imaging in psychophysiology: Potentialities and limits. *Psychophysiology,* 51, 951–963. Wiley Periodicals, Inc. Printed in the USA. DOI: doi.org/10.1111/psyp.12243.

ISO/IEC Guide 71 (2001). *Guidelines for Standards Developers to Address the Needs of Older Persons and Persons with Disabilities.* Geneva, Switzerland: ISO: International Organization for Standardization.

ISO/TC 173: Assistive Products (2019). ISO: International Organization for Standardization. Retrieved from https://www.iso.org/committee/53782/x/catalogue/. Accessed on: October 17th 2019.

ISO 9999 (2016). *Assistive Products for Persons with Disability: Classification and Terminology.* Geneva, Switzerland: ISO: International Organization for Standardization.

ISO 10377 (2013). *Consumer Product Safety: Guidelines for Suppliers.* Geneva, Switzerland: ISO: International Organization for Standardization.

ISO (2019a). *Standards Catalogue.* ISO: International Organization for Standardization. Retrieved from https://www.iso.org/ics/11.180.99/x/. Accessed on: July 4th 2019.

ISO 9241-210:2019 (2019b). *Ergonomics of Human-System Interaction: Part 210: Human-Centred Design for Interactive Systems.* Retrieved from https://www.iso.org/standard/77520.html. Accessed on: October 17th 2019.

ISO 9241-11:2018 (en) (2019c). *Ergonomics of Human-System Interaction: Part 11: Usability: Definitions and Concepts.* ISO Online Browsing Plataform (OBP). Retrieved from https://www.iso.org/obp/ui/#iso:std:iso:9241:-11:ed-2:v1:en. Accessed on: September 8th 2019.

Jacob, R.J.K.; Karn, K.S. (2003). Commentary on section 4. Eye tracking in human-computer interaction and usability research: Ready to deliver the promises. In: Radach, R.; Hyona, J.; Deubel, H. (eds.). *The Mind's Eye: Cognitive and Applied Aspects of Eye Movement.* Amsterdam: Elsevier, 573–605. ISBN: 9780444510204.

Japan Human Factors and Ergonomics Society (2020). Seven Practical Tips for Teleworking/ Home-Learning using Tablet/Smartphone Devices. União Latino-Americana de Ergonomia. IEA Press. Available at: https://secureservercdn.net/50.62.194.59/m4v.211 .myftpupload.com/wp-content/uploads/2014/10/7tips_guideline_0506_en_final.pdf Accessed on May15th 2021

Jenkins, D.W.; Davies, B.T. (1989). Product safety in Great Britain and the Consumer Protection Act 1987. *Applied Ergonomics*, 20, 213–217. DOI: 10.1016/0003-6870(89)90079-3.

Jenkins, S.; Brown, R.; Rutterford, N. (2009). Comparing thermographic, EEG, and subjective measures of affective experience during simulated interactions. *International Journal of Design*, 3(2), 53–65.

Jenkins, S.D.; Brown, R.D.; Donne, K.E. (2007). Infrared thermography in design research: The application of thermal imaging as a measurement tool in the design process. In P. Stebbing, G. Burden & L. Anusionwu (eds.). *Cumulus Working Papers: Schwäbisch Gmünd 18/07*. Helsinki: University of Art & Design Helsinki, 41–47.

Johnson, J. (2020). *Negative Effects of Technology: What to Know*. Medical News Today. Available at: https://www.medicalnewstoday.com/articles/negative-effects-of-technolo gy Accessed on May 18th. 2021.

Jones, J.C. (1992). *Design Methods*, 2nd ed. New York: Van Nostrand Reinhold. ISBN-13: 978-0471284963.

Jordan, P.W. (1998a). *An Introduction to Usability*. London: Taylor & Francis. ISBN-13: 978-0748407620.

Jordan, P.W. (1998b). Human factors for pleasure in product use. *Applied Ergonomics*, 29(1), 25–33.

Jordan, P.W. (2002). *Designing Pleasurable Products*. London: Routledge. ISBN-13: 978-0415298872.

Juran, J.M. (1992). *Juran on Quality by Design: The New Steps for Planning Quality into Goods and Services*. New York: Free Press. SBN-13: 978–0029166833.

Juran, J.M. (2016). *Juran's Quality Handbook: The Complete Guide to Performance Excellence*. 7th Ed. New York: McGraw-Hill Education.

Kahneman, D. et al. (2004). A survey method for characterizing daily life experience: The day reconstruction method. *Science*, 306(5702), 1776–1780. DOI: 10.1126/science. 1103572.

Karapanos, E.; Zimmerman, J.; Forlizzi, J.B. (2009). User experience over time: An initial framework. In: CHI'09: Proceedings of the 27th International Conference on Human Factors in Computing Systems. ACM, 729–738. DOI: 10.1145/1518701.1518814.

Karwowski, W.; Noy, Y. (2005). *Handbook of Human Factors in Litigation*. Boca Raton, FL: CRC Press. ISBN-13: 978-0415288705.

Karwowski, W.; Soares, M.M.; Stanton, N. (2011a). *Human Factors and Ergonomics in Consumer Product Design: Methods and Techniques*. Boca Raton, FL: CRC Press. ISBN-13: 978-1420046281.

Karwowski, W.; Soares, M.M.; Stanton, N. (2011b). *Human Factors and Ergonomics in Consumer Product Design: Uses and Applications*. Boca Raton, FL: CRC Press. ISBN-13: 978-1420046243.

King, R. (1987). *Better Design in Half the Time, Implementing Quality Function Deployment in America*, 3rd ed., Goal Q.P.C. Inc. ISBN-13: 978-1879364011.

King, R. (1989). *Better Design in Half the Time, Implementing Quality Function Deployment in America*. 3a. ed. Goal Q P C In. ISBN-13: 978-1879364011.

King, et al. (2020). Problematic online gaming and the COVID-19 pandemic. *Journal of Behavioral Addictions*, 9(2), 184–186. doi: 10.1556/2006.2020.00016

Király, O. et al. (2020). Preventing problematic internet use during the COVID-19 pandemic: Consensus guidance. *Comprehensive Psychiatry*, doi: 10.1016/j. comppsych.2020.152180.

Kirwan, B. (1992a). Human error identification in human reliability assessment. Part 1: Overview of approaches. *Applied Ergonomics*, 23, 299–318.

_____ (1992b). Human error identification in human reliability assessment. Part 2: Detailed comparison of techniques. *Applied Ergonomics*, 23, 371–381.

Kirwan, B.; Ainsworth, L.K. (1993). *A Guide to Task Analysis*. London: Taylor & Francis. ISBN 13: 978074840057.

Komninos, A. (2020). Norman's Three Levels of Design. In: *Interaction Design*. Retrieved from https://www.interaction-design.org/literature/article/norman-s-three-levels-of-d esign. Accessed on: August 19th 2020.

Kotler, P.; Armstrong, G. (2018). *Principles of Marketing*, 17th ed. London: Peason. ISBN-13: 978-0134492513.

Kramer, A.F.; McCarley, J.S. (2003). Oculomotor behaviour as a reflection of attention and memory processes: Neural mechanisms and applications to human factors. *Theoretical Issues in Ergonomics Science*, 4(1–2), 21–55. DOI: 10.1080/1463922 0210159744.

Kreifeldt, J. (2007). Ergonomics of product design. In: Salvendy, G. (ed.). *Handbook of Industrial Engineering*, 3rd. ed. New York: Wiley. ISBN-13: 978-0471502760.

Kreifeldt, J.; Alpert, M. (1985), Use, misuse, warnings: A guide for design and the law. In: Valseth, T.O. (ed.). Interface 85. Proceedings of the Fourth Symposium on Human Factors and Industrial Design in Consumer Products. Santa Monica, CA, Human Factors Society, 77–82.

Kroemer, K.H.E (2017). *Fitting the Human: Introduction to Ergonomics/Human Factors Engineering*, 7th ed. Boca Raton, FL: CRC Press. ISBN-13: 978-1498746892.

Kroemer, K.H.E.; Grandjean, E. (2004). *Ergonomics Handbook: Adapting the work to the man (Manual de Ergonomia: Adaptando o trabalho ao homem)*. In Portuguese. 5th ed. Porto Alegre: Bookman. ISBN-13: 978-8536304373.

Kroemer, K.H.F.; Kroemer, H.B. and Kroemer-Elbert, K.E. (2018). *Ergonomics: How to Design for Ease and Efficiency*, 3rd ed. Cambridge, MA: Academic Press. ISBN-13: 978-0128132968.

Krueger, R.A.; Casey, M.A. (2014). *Focus Groups: A Practical Guide for Applied Research*, 5th ed. ISBN-13: 978-1483365244.

Kujala, S. et al. (2011). UX Curve: A method for evaluating long-term user experience. *Interacting with Computers*, 23(5), 473–483. DOI: 10.1016/j.intcom.2011.06.005.

Kumar, S. (2009). *Ergonomics for Rehabilitation Professionals*. Boca Raton, FL: CRC Press. ISBN-13: 978-0849381461.

Kumar, S. (2007). *Perspectives in Rehabilitation Ergonomics*. London: Taylor and Francis. ISBN-13: 978-0748406739.

Langford, J.; McDonagh, D. (2003). *Focus Groups: Supporting Effective Product Development* (ed.). London: Taylor and Francis.

Laughery, K.R. (1993). Everybody knows - or do they? *Ergonomics in Design*, July, 8–13.

Leahy, J.A. (2021). Targeted focus groups in product development. Center on Knowledge Translation for Technology Transfer. Available at: http://publichealth.buffalo.edu/ content/dam/sphhp/cat/kt4tt/pdf/targeted-focus-groups-in-product-development.pdf. Accessed on: September 8th 2021.

Lee, S.H.; Harada, A.; Stappers, P.J. (2002), Pleasure with products: Design based on Kansei. In: Green, W. and Jordan, P. (eds.). *Pleasure with Products: Beyond Usability*. London: Taylor & Francis, 219–229. ISBN-13: 978-0415237048.

Legislation.gov.uk (2019). *Consumer Protection Act*. Retrieved from https://www.legislation .gov.uk/ukpga/1987/43. Accessed on January 2nd 2021.

Leonard, S.D.; Digby, S.E. (2003). Consumer perceptions of safety of consumer products. In: Kumar, S. (ed.). *Advances in Industrial Ergonomics and Safety IV*. London: Taylor & Francis, 169–176. DOI: 10.1201/9781482272383.

Leventhal, L.; Barnes, J. (2007). *Usability Engineering: Process, Products & Examples.* London: Pearson, ISBN-13: 978-0131570085.

Li, C.H.; Lau, H.K. (2018). Integration of industry 4.0 and assessment model for product safety. *2018 IEEE Symposium on Product Compliance Engineering (ISPCE),* 1–5 DOI: 10.1109/ISPCE.2018.8379269.

Li, Q.; Karreman, J.; Jong, M. (2018). *Chinese Technical Communicators' Opinions on Cultural Differences between Chinese and Western User Manuals.* IEEE Xplore, Digital Library. Retrieved from https://ieeexplore.ieee.org/abstract/document/8804552. Accessed on: September 23rd 2019.

Lidwell, W.; Holden, K.; Butler, J. (2010). *Universal Principles of Design.* Beverly, MA: Rockport Publisher, ISBN-13: 978-1592535873.

Lingaard, G. (1989). Defining what helps: An iterative approach the systems design. In: Proceeding of the 25th Annual Conference of the Ergonomics Society of Australia. Canberra, Australia.

Löbach, B. (2001). *Design Industrial: Bases Para a Configuração dos Produtos Industriais.* São Paulo: Edgard Blucher.

Lokman, A.M. (2010). Design & emotion: The Kansei engineering methodology. *Malaysian Journal of Computing,* 1(1), 1–14. Available at: https://anitawati.uitm.edu.my/mypapers/21_MJOC10_Design&Emotion_TheKEMethodology.pdf. Accessed on May 20th 2021.

Luchs, M.; Swan, S.; Griffin, A. (2015). *Design Thinking: New Product Development Essentials from the PDMA.* New York: Wiley-Blackwell. ISBN-13: 978-1118971802.

Lueder, R.; Rice, V.J.B. (2008). *Ergonomics for Children: Designing Products and Places for Toddlers to Teen.* London: Taylor & Francis. ISBN: 0415304741.

LUMA Institute (2012). *Innovating for People Handbook of Human-Centered Design Methods.* Pittsburgh, PA: LUMA Institute. ISBN-13: 978-0985750909.

Malmivuo, J.; Plonsey, R. (1995). *Bioelectromagnetism: Principles and Applications of Bioelectric and Biomagnetic Fields.* Oxford: Oxford University Press. ISBN-13: 978-0195058239.

Mano, H.; Oliver, R.L. (1993). Assessing the dimensionality and structure of the consumption experience: Evaluation, feeling, and satisfaction. *Journal of Consumer Research,* 20, December, 451–466.

Melo, C. (2020). Como o coronavírus vai mudar nossas vidas: dez tendências para o mundo pós-pandemia (How the coronavirus will change our lives: Ten trends for the post-pandemic world). In Portuguese. El País. Available at: https://brasil.elpais.com/opiniao/2020-04-13/como-o-coronavirus-vai-mudar-nossas-vidas-dez-tendencias-para-o-mundo-pos-pandemia.html Accessed on May 14th. 2021.

MentalHealth.gov (2021). What is mental health? Available at: https://www.mentalhealth.gov/basics/what-is-mental-health. Accessed on: August 25th 2021.

Merizi et al. (2018). *Methods for evaluating user experience in product design.* (Métodos para avaliação de experiência do usuário no design de produtos) In Portuguese. *Human Factors in Design,* 7(14), 114–132. DOI: 10.5965/2316796307142018114.

Miaskiewicz, T.; Kozar, K.A. (2011). Personas and user-centered design: How can personas benefit product design processes? *Design Studies,* 32(5), 417–430. DOI:10.1016/j.destud.2011.03.003.

Mace, R.; Hardie G.; Plaice J. (1991). Accessible environments: Toward universal design. In W. Preiser et al. (eds.). *Design Interventions: Toward a More Humane Architecture.* New York: Van Nostrand Reinhold.

Maldonado, T. (1977). *El diseño industrial reconsiderado.* Barcelona: Gustavo Gili. (em Espanhol).

Magrab, E.B. (2009). *Integrated Product and Process Design and Development: The Product Realization Process.* Boca Raton, FL: CRC Press. ISBN 9781420070606.

Maguire, M.C. (2001) Methods to support human-centred design, *International Journal of Human-Computer Studies*, 55, 4. DOI: 10.1006/ijhc.2001.0503.

Marconi, M.A.; Lakatos, E.M. (2017). *Fundamentals of Scientific Methodology (Fundamentos de Metodologia Científica)*. In Portuguese. São Paulo: Atlas. ISBN-13: 978-8597010121.

Marçal, M.A.; Silva, F.F.D.; Neto, L.F.M. (2016). Infrared thermography: Evaluation of skeletal muscle overload in the lumbar region and lower limbs in a production line (Termografia Infravermelha: Avaliação da Sobrecarga músculo Esquelética na Região lombar e Membros Inferiores em uma Linha de Produção). In: Portugues 5th Latin American Congress and 4th Peruvian Congress of Ergonomics. Lima, Peru.

Margolin, V.; Buchanan, R. (1995). *The Idea of Design*. Cambridge, MA: MIT Press. ISBN-13: 978-0262631662.

Maritan, D. (2015). *Practical Manual of Quality Function Deployment*. Cham: Springer. ISBN-13: 978-3319085203.

Mash, J. (2016). *UX for Beginners: A Crash Course in 100 Short Lessons*. 'Reilly Media. ISBN-13: 978-1491912683.

McKey, E. (2013). *UI is Communication: How to Design Intuitive, User Centered Interfaces by Focusing on Effective Communication*. Waltham, MA: Morgan Kaufmann. ISBN-13: 978-0123969804.

Menon, U.; O'Grady, J.; Gu, J.Z.; Young, E. (1994). Quality function deployment: An overview. In: Syan, C.S. and Menon, U. (eds.). *Concurrent Engineering: Concepts, Implementation and Practice*. London: Chapman & Hall.

Merla, A.; Romani, G.L. (2007). Thermal signatures of emotional arousal: A functional infrared imaging study. In: Proceedings of the 29th Annual International Conference of the IEEE EMBS Cité Internationale. Lyon, France, August 23–26, 2007.

Milton, A.; Rodgers, P. (2013). *Research Methods for Product Design*. London: Laurence King Publishing. ISBN-13: 978-1780673028.

Mital, A. (1995). The role of ergonomics in designing for manufacturability and humans in general in advanced manufacturing technology: Preparing the American workforce for global competition beyond the year 2000. *International Journal of Industrial Ergonomics*, 15(2), February, 129–135. https://doi.org/10.1016/0169-8141(94)00073-C

Mital, A.; Anand, S. (1992). Concurrent design of products and ergonomic considerations. *Journal of Design and Manufacturing*, 2, 167–183.

Mital, A.; Morse, I.E. (1992). The role of ergonomics in designing for manufacturability. In: Helander, M. and Nagamachi, M.(eds.). *Design for Manufacturability: A System Approach for Concurrent Engineering and Ergonomics*. London: Taylor & Francis, 147–159.

Mitchell, J. (1981). User requirements and the development of products which are suitable for the broad spectrum of user capacities. *Ergonomics*, 24(11), 863–869. DOI: 10.1080/00140138108924906.

Møller, M.H. (2013). Usability testing of user manuals. *Communication & Language at Work*, 2(2), 51. DOI: 10.7146/claw.v1i2.7892.

Moraes, A. (1992). *Ergonomic Diagnosis of the Communicational Process of the Man-Machine Data Transcription System: Data Entry Workstation in Informational Data Entry Terminals (Diagnóstico ergonômico do processo comunicacional do sistema homem-máquina de transcrição de dados: posto de trabalho do digitador em terminais informativos de entrada de dados)*. In Portuguese, 4 vols. Ph.D. Thesis. Federal University of Rio de Janeiro, Communication School, Rio de Janeiro, Brazil.

Moraes, A. (2013). Ergonomics, Ergodesign and Usability: Some stories, precursors; divergences and convergences (Ergonomia, Ergodesign e Usabilidade: algumas histórias, precursosres; divergências e convergências). In Portuguese. *Revista Ergodesign HCI*, 1(1), 1–9, Retrieved from http://periodicos.puc-rio.br/index.php/revistaergodesign-hci/article/view/41. Accessed on: September 8th 2020. DOI: 10.22570/ergodesignhci.v1i1.41.

Moraes, A.; Mont'Alvão, C. (2010). *Ergonomics: Concepts and Applications (Ergonomia: Conceitos e Aplicações)*, 4th ed. Rio de Janeiro: 2AB. ISBN-13: 978-8586695490.

Mossel, W.P.; Christiaans, H.H.C.M. (1991). Designers and the handling of their products. In: Lovesey, E.S. (ed.). Contemporary Ergonomics 1991. Proceedings of the Ergonomics Society's 1991 Annual Conference. London, Taylor & Francis, 370–374.

Mowen, J.C.; Minor, M. (1997). *Consumer Behavior*, 5th ed. London: Prentice Hall. ISBN-10: 0137371152.

Nagamashi, M. (2016a). Home applications of Kansei engineering in Japan: An overview. *Gerontechnology*, 15(4), 209–215. DOI: 10.4017/gt.2016.15.4.005.00.

Nagamashi, M. (2017). *Innovations of Kansei Engineering*. Boca Raton, FL: CRC Press. ISBN-13: 978-1138440609.

Nagamashi, M. (2016b). *Kansei/Affective Engineering*. Boca Raton, FL: CRC Press. ISBN-13: 978-1138440593

Nagamashi, M. (2002). Kansei Engineering as an ergonomic consumer-oriented technology for product development. *Applied Ergonomics*, 33, 289–294. DOI: 10.1016/0169-8141(94)00052-5.

National Disability Authority (NDA) (2020). *The 7 Principles*. Retrieved from http://universa ldesign.ie/What-is-Universal-Design/The-7-Principles/#p1. Accessed on: November 10th 2020.

Nichols, P.J.R. (1976). Aids for daily living: The problems of the severely disabled. *Applied Ergonomics*, 7(3), 126–132.

Nielsen, J. (1993). *Usability Engineering*. Cambridge, MA: Academic Press. ISBN-13: 978-0125184069.

Nielsen Norman Group (2019a). *How Many Test Users in a Usability Study?* Retrieved from https://www.nngroup.com/articles/how-many-test-users/ Accessed on: September 16th 2019.

Nielsen Norman Group (2020). *Introduction to Usability*. Retrieved from https://www.nngroup.com/articles/usability-101-introduction-to-usability/. Accessed on: August, 12th 2020.

Nielsen Norman Group (2019b). *Why You Only Need to Test with 5 Users*. Retrieved from https://www.nngroup.com/articles/why-you-only-need-to-test-with-5-users/. Accessed on: September 23rd 2019.

Noor, Y.M. et al. (2019). Investigating the product quality attributes that influence customers satisfaction of online apparels. *International Journal of Advance Research*, 7(20), 819–827. DOI: 10.21474/IJAR01/8551.

Norman, D. (2008). *Design emocional*. In Portuguese. Rio de Janeiro: Rocco. ISBN-13: 978-8532523327.

Norman, D.A. (2013). *The Everyday Design (O design do dia a dia)*. In Portuguese. Anfiteatro. Edição do Kindle.

Norman, D.A. (1988). *The Psychology of Everyday Things*. New York: Basic Books. ISBN-13: 978-0465067091.

Noy, Y.I.; Karwowski, W. (2004). *Handbook of Human Factors in Litigation*. Boca Raton, FL: CRC Press. ISBN-13: 978-0415288705.

Office of Financial Management (2021). Telework resources during the COVID-19 pandemic. Washington State Office of Financial Management. Available at: https://ofm.wa.gov /state-human-resources/coronavirus-covid-19-hr-guidance-state-agencies/telework-re sources-during-covid-19-pandemic. Accessed on: May 22nd 2021.

Ohkura, M. (2019). *Kawaii Engineering: Measurements, Evaluations, and Applications of Attractiveness*. Cham: Springer. ISBN-13: 978-9811379635

Oliver, R.L. (1993). Cognitive, affective, and attribute bases of the satisfaction response. *Journal of Consumer Research*, 20, December, 418–430.

Olsen, D. (2015). *The Lean Product Playbook: How to Innovate with Minimum Viable Products and Rapid Customer Feedback*. New York: Wiley. ISBN-13: 978-1118960875.

Onwubolu, G. (2013). *Computer-Aided Engineering Design with Solidworks*. London: Imperial College Press. ISBN-13: 978-1848166653.

Ottley, B.; Lasso, R.; Klely, T. (2013). *Understanding Products Liability Law*, 2nd ed. LexisNexis. ISBN-13: 978-0769863757.

Owen, D.; Davis, M. (2019). *Products Liability and Safety, Cases and Materials*, 7th ed. Minnesota: West Academic Press. ISBN-13: 978-1642420944.

PAHO (2021). COVID-19 factsheets: Understanding the infodemic and misinformation in the fight against COVID-19. Pan American Health Organization. Available at: https://iris.paho.org/handle/10665.2/52052. Accessed on September 9th 2021.

Panero, J.; Zelnik, M. (2016). *Human dimensioning for interior spaces: A Reference and Reference Book for Projects (Dimensionamento humano para espaços interiores: Um Livro de Consulta e Referência Para Projetos)*, 2nd ed. In Portuguese. Barcelona: Gustavo Gili ISBN-13: 978-8584520114.

Pannafino, J.; McNeil, P. (2017). *UX Methods: A Quick Guide to User Experience Research Methods*. CDUXP LLC. ISBN-13: 978-0692972717.

Parasuraman, R.; Rizzo, M. (2008). *Neuroergonomics: The Brain at Work*. Oxford: Oxford University Press. ISBN-13: 978-0195368659.

Pavel, N.; Zitkus, M. (2017). Extending product affordances to user manuals. In: The 9th International Conference on Engineering and Product Design Education. September, 7th and 8th, 2017. Norway, Oslo and Akershus University of College of Applied Sciences. Retrieved from https://oda-hioa.archive.knowledgearc.net/bitstream/handle/10642/6069/extending%2bproduct%2baffordances%2bto%2bmanuals.pdf?sequence=1&isAllowed=y.

Pavliscak, P. (2018). *Emotionally Intelligent Design: Rethinking How We Create Products*. Sebastopol, Canada: O'Reilly Media. ISBN-13: 978-1491953143.

Pine, B.J.; Gilmore, J.H. (2020). *The Experience Economy: Competing for Customer Time, Attention, and Money*. Brighton: Harvard Business Review Press.

Persson, H.; Åhman, H.; Yngling, A.A. et al. (2015). Universal design, inclusive design, accessible design, design for all: Different concepts—one goal? On the concept of accessibility—historical, methodological and philosophical aspects. *Universal Access in the Information Society*, 14, 505–526. DOI: 10.1007/s10209-014-0358-z

Peters, G.A.; Peters, B.J. (2006). *Human Error: Causes and Control*. Boca Raton, FL: CRC Press. ISBN-13: 978-0849382130.

Pheasant, S.; Haslegrave, C. (2018). *Bodyspace: Anthropometry, Ergonomics and the Design of Work*, 3rd ed. London: Taylor & Francis. ISBN-13: 978-0415285209.

Pirkl, J.J. (1994). *Transgenaration Design: Products for an Aging Population*. New York: Van Nostrand Reinhold. ISBN-13: 978-0442010652.

Poole, A.; Ball, L.J. (2005). Eye tracking in human-computer interaction and usability research: Current status and future prospects. In: Ghaoui, C. (ed.). *Encyclopedia of Human Computer Interaction*. New York: Idea Group Reference, 211–219. ISBN-13: 978-1591405627.

Poulson, D.; Ashby, M.; Richardson, S. (1996). *UserFit: A Practical Handbook on User-Centred Design for Assistive Technology*. Loughborough, UK: HUSAT Research Institute.

Potas, N.; Açıkalın, Ş.N., Erçetin, Ş.Ş. et al. (2021). Technology addiction of adolescents in the COVID-19 era: Mediating effect of attitude on awareness and behavior. *Current Psychology* https://doi.org/10.1007/s12144-021-01470-8

Pressman, A. (2018). *Design Thinking: A Guide to Creative Problem Solving for Everyone*. London: Routledge. ISBN-13: 978-1138673472.

Privitera, M.B. (2019). *Applied Human Factors in Medical Device Design.* Cambridge, MA: Academic Press. ISBN-13: 978-0128161630.

Product Life Cycle Stages (2019). *Product Life Cycle Stages.* Retrieved from http://product lifecyclestages.com. Accessed on: June 28th 2019.

Pugh, S. (1991). *Total Design: Integrated Methods for Successful Product Engineering.* Boston, MA: Addison-Weley. ISBN-10: 0201416395.

Pullin, G. (2011). *Design Meets Disability.* Cambridge, MA: The MIT Press. ISBN-13: 978-0262516747.

Quaresma, M. et al. (2021). UX concepts and perspectives – From usability to user experience design. In Soares, M.M.; Rebelo, F.; Ahram, T. (eds.). *Handbook of Usability and User Experience: Research and Case Studies.* Boca Raton: CRC Press.

Rahayu, F.S. et al. (2020). Research trend on the use of IT in digital addiction: An investigation using a systematic literature review. *Future Internet,* 12, 174. doi: 10.3390/fi12100174.

Reason, J. (1990). *Human Error.* New York: Cambridge University Press. ISBN-13: 978-0521314190.

Rebelo, F.; Duarte, E.; Noriega, P.; Soares, M. (2011). Virtual reality in consumer product: Design, methods and applications. In: Karwowski, W.; Soares, M.M.; Stanton, N.A. (Org.). *Human Factors and Ergonomics in Consumer Product Design: Methods and Techniques,* vol. 1. Boca Raton, FL: CRC Press, 381–404. ISBN-13: 978-142004 6281.

Rebelo, F. et al. (2021). Advanced user experience evaluations using biosensors in virtual environments. In Soares, M.M.; Rebelo, F.; Ahram, T. (eds.). *Handbook of Usability and User Experience: Methods and Techniques.* Boca Raton: CRC Press.

Rebelo, F.; Noriega, P.; Duarte, E.; Soares, M. (2012). Using virtual reality to access user experience. *Human Factors,* 54, 964–982. DOI: 10.1177/0018720812465006.

Reiss, E. (2012). *Usable Usability: Simple Steps for Making Stuff Better.* New York: John Wiley & Sons.

Research Institute for Consumer Affairs (1984). *Aids for People with Disabilities; Bibliography with Summaries of Performance Studies.* London, UK, 933 pp.

Research Institute for Disabled Customers (2020). Digital inclusion. The experience of disabled people during the Covid-19 pandemic restrictions shows technology as a powerful amplifier. Available at: https://www.ridc.org.uk/content/research-and-consultancy/our-insights/covid-19-stories/digital-inclusion. Accessed on: August 25th 2021.

Robert, A.; Roth, S.; Chamoret, D.; Yan, X.; Peyraut, F.; Gomes, S. (2012). Functional design method for improving safety and ergonomics of mechanical products. *Journal of Biomedical Science and Engineering,* 5, 457–468. DOI: 10.4236/jbise.2012.58058.

Roebuck, J.A. (1995). *Anthropometric Methods: Designing to Fit the Human Body.* Santa Monica, CA: Human Factors and Ergonomics Society. ISBN-13: 978-0945289012.

Roebuck Jr., J.A.; Kroemer, K.H.E.; Thomson, W.G. (1995). *Anthropometric Methods: Designing to Fit the Human Body.* Santa Monica, CA: Human Factors and Ergonomics Society. ISBN-13: 978-047172975.

ROSPA (2020). *Accident Statistics.* The Royal Society for the Prevention of Accidents. Available at: https://www.rospa.com/resources/statistics/. Accessed on: June 10th 2020.

Rosson, M. E; Carroll, J. (2001). *Usability Engineering: Scenario-based Development of Human-Computer Interaction.* San Francisco, CA: Morgan Kaufmann. ISBN-13: 978-1558607125.

Roy, R., 2018. Consumer product design: Patterns of innovation, market success and sustainability. *Journal of International Business Research and Marketing,* 3(5), 25–33. DOI: 10.18775/jibrm.1849-8558.2015.35.3004.

Rozenburg, N.F.M.; Eekels, J. (1995). *Product Design: Fundamentals and Methods.* Chichester and New York: Wiley. ISBN-13: 978-0471954651.

Rubin, J.; Chisnell, D. (2008). *Handbook of Usability Testing: How to Plan, Design, and Conduct Effective Tests.* New York: Wiley. ISBN-13: 978–0470185483.

Ryan, J.P. (1987). Consumer behaviour considerations in product design. In: Proceedings of the Human Factors Society: 31st Annual Meeting. Santa Monica, CA, Human Factors Society, 1236–1239.

Ryan, J.P. (1985). Do safety standards make safe products? In: Kvalseth, T.O. (ed.). Interface 85. Proceedings of the Fourth Symposium on Human Factors and Industrial Design in Consumer Products. Santa Monica, CA, Human Factors Society, 119–124.

Sacharin, V.; Schlegel, K.; Scherer, K.R. (2012). *Geneva Emotion Wheel Rating Study* (Report). Geneva, Switzerland: University of Geneva, Swiss Center for Affective Sciences.

Sadeghi, L.; Dantan, Y.; Mathieu, L.; Siadat, A.; Aghelinejad, M.M. (2017). A design approach for safety based on product-service systems and function–behavior–structure. *CIRP Journal of Manufacturing Science and Technology*, 19, 44–56. DOI: 10.1016/j.cirpj.2017.05.001.

Salvendy, G. (2012). *Handbook of Human Factors and Ergonomics*, 4th ed. New York: Wiley. ISBN-13: 978-0470528389.

Sanders, M.S.; McCormick, E.J. (1993). *Human Factors in Engineering and Design*, 7th ed. New York: McGraw-Hill. ISBN-13: 978-0070549012.

Sangelkar, M.; Mcadams, D. (2012). Adapting ADA architectural design knowledge for universal product design using association rule mining: A function based approach. *Journal of Mechanical Design*, 134(7), 1–15. DOI: 10.1115/1.4006738.

Santos, M.S.E.; Soares, M.M. (2016). Ergonomic design thinking? A project management model for workplace design. In: Soares, M.M.; Rebelo, F. (Org.). *Ergonomics in Design: Methods and Techniques*, vol. 1, 1st ed. Boca Raton, FL: Estados Unidos, CRC Press, 267–280. ISBN-13: 978-1498760706.

Schumacher, H. (2019). Inside the world of instruction manuals. *Follow BBC Future.* Disponível em: http://www.bbc.com/future/story/20180403-inside-the-world-of-instruction-manuals. Accessed on: September 23rd 2019.

Schütte, S.T.W.; Eklund, J.; Axelsson, J.R.C.; Nagamachi, M. (2004). Concepts, methods and tools in Kansei engineering. *Theoretical Issues in Ergonomics Science*, 5(3), 214–231. DOI: 10.1080/1463922021000049980.

Seva, R.; Seva, R.; Gosiaco, K.; Santos, M.C.; Pangilinan, D. (2011). Product design enhancement using apparent usability and affective quality. *Applied Ergonomics*, 42, 511–517. DOI: 10.1016/j.apergo.2010.09.009.

Shackel, B. (2009). Usability: Context, framework, definition, design and evaluation. *Interacting with Computers*, 21(5–6), 339–346. DOI: 10.1016/j.intcom.2009.04.007.

Shin, D.; Wang, Z. (2015). The experimentation of matrix for product emotion. *Procedia Manufacturing*, 3, 2295–2302. In: 6th International Conference on Applied Human Factors and Ergonomics (AHFE 2015) and the Affiliated Conferences, AHFE 2015. DOI: 10.1016/j.promfg.2015.07.375.

Shorrock, S.; Williams, C. (2016). *Human Factors and Ergonomics in Practice.* London: Routledge. ISBN-13: 978-1472439253.

Singh, A.K.; Singh, P.K. (2019). Digital addiction: A conceptual overview. *Library Philosophy and Practice (e-journal).* 3538. https://digitalcommons.unl.edu/libphilprac/3538

Smith, I. (2003). *Meeting Customer Needs*, 3rd ed. Oxford, UK: Butterworth Heinemann.

Smith, I. (1987). The case of the missing human factors data. In: Proceedings of the Human Factors Society: 31st Annual Meeting. Santa Monica, CA: Human Factors Society, 1042–1043.

Soares, M.M. (1990). *Human costs for sitting posture and parameters for the evaluation and design of "university desk chairs": A case study (Custos humanos para a postura sentada e parâmetros para a avaliação e design de "carteiras universitárias": um estudo de caso).* In Portuguese. M.Sc. Thesis. Federal University of Rio de Janeiro, COPPE, Brazil.

Soares, M.M. (1999). *Translating User Needs into Product Design for Disabled People: A Study of Wheelchairs.* Ph.D. Thesis. Loughborough University, UK.

Soares, M.M. (2012). Translating user needs into product design for the disabled: An ergonomic approach. *Theoretical Issues in Ergonomics Science*, 13, 92–120. https://doi.org/10.1080/1463922X.2010.512989

Soares, M.M.; Bucich, C.C. (2000). Product safety: Reducing accidents through design (Segurança do produto: reduzindo acidentes através do design). In Portuguese. *Estudos em Design, Rio de Janeiro*, 8(2), 43–67.

Soares, M.M.; Rebelo, F. (2017). *Ergonomics in Design: Methods and Techniques.* Boca Raton, FL: CRC Press. ISBN-13: 978-1498760706.

Soares, M.M.; Rebelo, F.; Ahram, T. (2021). *Handbook of Usability and User Experience (UX)*, 2 vols. Boca Raton, FL: CRC Press. ISBN: 9780367357689.

Soares, M.M. et al. (2021). Usability and user experience: Methods and models. In Soares, M.M.; Rebelo, F.; Ahram, T. (eds.). *Handbook of Usability and User Experience: Methods and Techniques.* Boca Raton: CRC Press.

Soares M.M.; Vitorino D.F.; Marçal M.A. (2019a) Application of digital infrared thermography for emotional evaluation: A study of the gestural interface applied to 3D modeling software. In: Rebelo F., Soares M. (eds.). *Advances in Ergonomics in Design. AHFE 2018. Advances in Intelligent Systems and Computing*, vol. 777. Cham: Springer. DOI: 10.1007/978-3-319-94706-8_23.

Soares, M.M.; Vitorino, D.F.; Marçal, M.A. (2019b). Application of digital infrared thermography for emotional evaluation: A study of the gestural interface applied to 3D modeling software. In: Rebelo, Francisco; Soares, Marcelo M. (Org.). *Advances in Intelligent Systems and Computing*, vol. 1, 1st ed. Cham: Springer, 201–212.

Soede, M. (1990). Rehabilitation technology or the ergonomics of ergonomics. *Ergonomics*, 33(3), 367–373. DOI: 10.1080/00140139008927138.

Solomon, M.R. (2016). *Consumer Behavior: Buying, Having, and Being*, 12nd ed. London: Pearson. ISBN-10: 9780134129938.

Stadler-Estrin, K.; Estrin, S.A. (1987). Consumer products: The failure to warn: How hazardous is it? In: Interface 87: Human Implication of Product Design. Proceeding of the 5th Symposium on Human Factors and Industrial Design in Consumer Products. Santa Monica, CA, Human Factors Society, 237–243.

Stearn, M.C.; Galer, I.A.R. (1990). Increasing consumer awareness: An ergonomics marketing strategy for the future. *Ergonomics*, 33, 341–347.

Still, B.; Crane, K. (2016). *Fundamentals of User-Centered Design: A Practical Approach.* Boca Raton, FL: CRC Press. ISBN-13: 978-1498764360.

Stuster, J. (2019). Task analysis: How to develop an understanding of work. Washington D.C.: Human Factors and Ergonomics Society. ISBN-13: 978-0945289579.

Sullivan, L.P. (1986). Quality function deployment. *Quality Progress*, June, 39–50.

Sun, X.; Houssin, R; Renaud, J.; Gardoni, M. (2018). Towards a human factors and ergonomics integration framework in the early product design phase: Function-task-behaviour. *International Journal of Production Research*, 56(14), 4941–4953. DOI: 10.1080/00207543.2018.1437287.

Sun, S.; Lin, D.; Operario, D. (2020). Need for a population health approach to understand and address psychosocial consequences of COVID-19. *Psychological Trauma: Theory, Research, Practice, and Policy*, 12(S1), S25. doi: 10.1037/tra0000618.

Suryadi, N; Suryana, R; Komaladewi, R.; Sari, D. (2018). Consumer, customer and perceived value: Past and present. *Academy of Strategic Management Journal*, 17(4), 1–9. ISBN: 1939-6104-17-4.

Swallow, E. (2018). *Product Development: How and Why You Should Include Your Customers*. Retrieved from https://www.weebly.com/inspiration/product-development-include-customers/. Accessed on: June10th 2019.

Swiss Center for Affective Sciences (2020). *The Geneva Emotion Wheel*. Université de Geneve. Retrieved from https://www.unige.ch/cisa/gew/. Accessed on: August 25th 2020.

Terminko, J. (1997). *Step-by-Step QFD: Customer-Driven Product Design*, 2nd ed. Boca Raton, FL: CRC Press. ISBN-13: 978-1574441109.

The World Bank (2020). *Disability Inclusion*. Retrieved from https://www.worldbank.org/en/topic/disability. Accessed on: July 16th 2020.

Thimbleby, H. (1991). Can humans think? The ergonomics society lecture 1991. *Ergonomics*, 34, 1269–1287.

Tilley, A.; Henry Dreyfuss Associates, (2001). *The Measure of Man and Woman: Human Factors in Design*. New York: John Wiley & Sons. ISBN-13: 978–0471099550.

Tillman, B.; Tillman, P.; Rose, R.R.; Woodson, W.E. (2016). *Human Factors and Ergonomics Design Handbook*, 3rd. ed. New York: McGraw-Hill Education. ISBN-13: 978-0071702874.

Torrens, G. (2011). Universal design: Emphaty and affinity. In: Karwowski, W.; Soares, M.M.; Stanton, N.A. (Org.). *Human Factors and Ergonomics in Consumer Product Design: Methods and Techniques*. Boca Raton, FL: CRC Press, vol. 1, p. 233–248. ISBN-13: 978-1420046281.

Tripathi, A. (2018). Impact of internet addiction on mental health: An integrative therapy is needed. Integrative Medicine. Available at: https://www.karger.com/Article/Pdf/491997. Accessed on May18th 2021

Tse, D.K.; Wilton, P. (1988). Models of consumer satisfaction formation: An extension. *Journal of Marketing Research*, May, 204–212. 25(2). DOI: 10.1177/002224378802500209.

Ulrich, K.T. and Eppinger, S.D. (1995). *Product Design and Development*. New York, McGraw-Hill.

Ulrich, K.T.; Eppinger, S.D. (2019). *Product Design and Development*, 7th ed. New York: McGraw-Hill. ISBN-13: 978-1260134445.

UNBC (2021). *Defeating Your Digital Dependency*. Academic Success Center. University of Northern British Columbia. Available at: https://www2.unbc.ca/sites/default/files/sections/academic-success-centre/digitaldependency.pdf. Accessed on: May18th 2021

United Nations Disability Statistics Database (2020). United Nations Statistics Division. Retrieved from https://unstats.un.org/unsd/demographic-social/sconcerns/disability/statistics/#/countries. Accessed on: July 15th 2020.

Vanderheiden, G.C. (1990). Thirty-something million: Should they be exceptions? *Human Factors*, 32(4), 383–396. DOI: 10.1177/001872089003200402.

Vanderheiden, G.C.; Jordan, J.B. (2012). Design for people with functional limitations. In: Salvendy, G. *Handbook of Human Factors and Ergonomics*, 4th ed. New York: Wiley, Chapter 51, p. 1409–1441. ISBN-13: 978-0470528389.

Vanderheiden, G.C.; Vanderheiden, K.R. (2019) *Accessible Design of Consumer Products: Guidelines for the Design of Consumer Products to Increase Their Accessibility to People with Disabilities or Who Are Aging*. Trace Research and Development Centre. Retrieved from https://trace.umd.edu/publications/consumer_product_guidelines. Accessed on: September 18th 2019.

Vanlandewijck, Y.C.; Spaepen, A.J.; Theisen, D. (2007). Mobility of the disabled: Manual wheelchair propulsion. In: Kumar, S. (ed.). *Perspectives in Rehabilitation Ergonomics*. London, Taylor and Francis.

Vermeeren, A.P.O.S. et al. (2010). User experience evaluation methods: Current state and development needs. In: *Proceedings of the 6th Nordic Conference on Human-Computer Interaction.* Reykjavik Iceland: Extending Boundaries. DOI: 10.1145/1868914.1868973.

Vincent, C.J.; Y, Li; Blandford, A. (2014). Integration of human factors and ergonomics during medical device design and development: It's all about communication. *Applied Ergonomics*, 45(3), 413–419. DOI: 10.1016/j.apergo.2013.05.009.

Virzi, R. (1992). Refining the test phase of usability evaluation: How many subjects is enough? *Human Factors*, 34, 457–468. DOI:10.1177/001872089203400407.

Vitorino, D.F. (2017). *Ergonomic and Usability Analysis with the Aid of Digital Infrared Thermography: A Study of the Gestural Interface Applied in 3D Modeling Software (Análise ergonômica e da usabilidade com auxílio da termografia digital por infravermelho: um estudo da interface gestual aplicada em software de modelagem 3D).* In Portuguese. M.Sc. Thesis. Post Graduate Program in Design, Federal University of Pernambuco, Brazil.

Vitorino, D.F. (2017). *Análise ergonômica e da usabilidade com auxílio da termografia digital por infravermelho: um estudo da interface gestual aplicada em software de modelagem 3D.* Dissertação de Mestrado. Programa de Pós-Graduação em Design, Universidade Federal de Pernambuco.

Walter, A. (2011). *Designing for Emotion.* A Book Apart, ISBN-13: 978–1937557003.

Wang, G.G. 2002. Definition and review of virtual prototyping. *Journal of Computing and Information Science in Engineering*, 2(3), 232–241. DOI: 10.1115/1.1526508.

Ward, S. (1992). Product design and ergonomics. *Ergonomics Australia*, 6, 15–18.

Ward, S. (1990). The designer as ergonomist, ergonomics design of products for the consumer. In: Proceedings of the 26th. Annual Conference of the Ergonomics Society of Australia. Branch, Kensington, South Australia, Ergonomics Society of Australia, 101–106.

Weiss, E.H. (1991). *How To Write Usable User Documentation*, 2nd ed. Greenwood. ISBN-13: 978-0897746397

Wendel, S. (2014). *Designing for Behavior Change.* O'Reilly Media. ISBN-13: 978-1449367626.

Whalen, J. (2019). Design for how people think. O'Reilly Media. ISBN-13: 978-1491985458.

What is the Difference between a Hazard and a Risk? Retrieved from https://hssewor ld.com/what-is-the-difference-between-a-hazard-and-a-risk/. Accessed on: July 20th. 2020.

Wheelchair Needs In The World (2016). Wheelchair Foundation. Retrieved from https://ww w.wheelchairfoundation.org/programs/from-the-heart-schools-program/materials-and -supplies/analysis-of-wheelchair-need/. Accessed on: June 10th. 2019.

WHO (2005). WHO – World Health Organization. Resource Book on Mental Health, Human Rights and Legislation. Genebra: World Health Organization.

WHO (2020). WHO – World Health Organization. Assistive Devices and Available at: https://www.who.int/disabilities/technology/en/. Accessed on: July 30th 2020.

WHO (2021a). Mental health: Strengthening our response. Available at: https://www.who.int/ news-room/fact-sheets/detail/mental-health-strengthening-our-response. Accessed on: May 14th 2021

WHO (2021b). Mental Health Considerations during COVID-19 Outbreak. Available at: https ://pscentre.org/wp-content/uploads/2020/03/WHO-mental-health-considerations.pdf. Accessed on: May 15th 2021

Wiese, B.S.; Sauer, J.; Rüttinger, B. (2004) Consumers' use of written product information, *Ergonomics*, 47(11), 1180–1194. DOI: 10.1080/00140130410001695951

Wilamowitz-Moellendorff, M.; Hassenzahl, M.; Platz, A. (2006). Dynamics of user experience: How the perceived quality of mobile phones changes over time. In: User

Experience: Towards a Unified View, Workshop at the 4th Nordic Conference on Human-Computer Interaction, 74–78.

Wilkoff, W.L.; Abed, L.W. (1994). *Practicing Universal Design: An Interpretation of the ADA*. New York: Van Nostrand Reinhold. ISBN-13: 978-0442013769.

Wilson, J.R. (1983). Pressures and procedures for the design of safer consumer products. *Applied Ergonomics*, 14, 109–116.

Wilson, J.R.; Kirk, N.S. (1980). Ergonomics and product liability. *Applied Ergonomics*, 11, 130–136. DOI: 10.1016/0003-6870(80)90001-0.

Wilson, J.R.; Rutherford, A. (1989). Mental models: Theory and application in human factors. *Human Factors*, 31, 617–634. DOI: 10.1177/001872088903100601.

Wilson, J.R.; Sharples, S. (2015). *Evaluation of Human Work: A Practical Ergonomics Methodology*, 4th. ed. London: Taylor & Francis. ISBN-13: 978-1466559615

Wood, D. (1990). Ergonomists in the design process. Contemporary products shed new light on avenues for interaction between ergonomists and industrial designers. In: Ergonomics Design of Products for the Consumer. Proceedings of the 26th. Annual Conference of the Ergonomics Society of Australia. Branch, Kensington, South Australia, Ergonomics Society of Australia, 125–132.

Woods, D.; Dekker, S.; Cook, R.; Johannesen, L. (2010). *Behind Human Error*, 2nd ed. London: Routledge. ISBN-13: 978-0754678342.

World Health Organization (WHO) (2020). *Assistive Devices and Technologies*. Retrieved from https://www.who.int/disabilities/technology/en/. Accessed on: July 30th 2020.

World Health Organization (WHO) (2005). *Resource Book on Mental Health, Human Rights and Legislation*. Genebra: World Health Organization.

Xiang, H.; Chany, A-M.; Smith, G.A. (2006). Wheelchair related injuries treated in US emergency departments. *Injury Prevention*, 12:8–11. DOI: 10.1136/ip.2005.010033.

Yalanska, M. (2021). FAQ: Human-centered vs user-centered. Are the terms different? Available at: https://blog.tubikstudio.com/faq-design-platform-human-centered-vs-user-centered-are-the-terms-different/. Accessed on: May 11th 2021.

Yen et al. (2008). Psychiatric symptoms in adolescents with Internet addiction: Comparison with substance use. *Psychiatry and Clinical Neurosciences*, 62, 9–16.

Yoshikawa, A. (2000). Subjective information processing: Its foundation and applications. *Biomedical Soft Computing and Human Sciences*, 6(1), 75–83. DOI: 10.24466/ijbschs.6.1_75.

Zairi, M. (1993). Quality function deployment: A modern competitive tool. *Technical Communications*. ISBN-13: 978-0946655724.

Zeng, J.; Soares, M.M.; He, R. (2020). Systematic review on using biofeedback (EEG and infrared thermography) to evaluate emotion and user perception acquired by Kansei engineering. In: Marcus A., Rosenzweig E. (Org.). *Lecture Notes in Computer Science*, vol. 12200, 1st ed. Cham: Springer, p. 582–593. DOI: 10.1007/978-3-030-49713-2_40.

Zhu, A.; Zedtwitz, M.; Assimakopoulos, A; Fernandes, K. (2016). The impact of organizational culture on concurrent engineering, design-for-safety, and product safety performance. *International Journal of Production Economics*, 176, 69–81. DOI: 10.1016/j.ijpe.2016.03.007.

Index